What Is Natural?

D0169763

Gift of the Knapp Foundation
2002-2003

Other Books by the Author

Beyond the Gene
Evolution by Association
Where the Truth Lies

What Is Natural?

Coral Reef Crisis

Jan Sapp

OXFORD

UNIVERSITY PRESS

OXFORD

UNIVERSITY PRESS

Oxford   New York

Auckland   Bangkok   Buenos Aires   Cape Town   Chennai
Dar es Salaam   Delhi   Hong Kong   Istanbul   Karachi
Kolkata   Kuala Lumpur   Madrid   Melbourne   Mexico City
Mumbai   Nairobi   São Paulo   Singapore   Taipei   Tokyo   Toronto

Copyright © 1999 by Jan Sapp

First published by Oxford University Press Inc., 1999
First issued as an Oxford University Press paperback, 2003
198 Madison Avenue, New York, New York 10016

www.oup.com

Oxford is a registered trademark of Oxford University Press

Library of Congress Cataloging-in-Publication Data
Sapp, Jan
What is natural? :
coral reef crisis / Jan Sapp.
p.   cm.   Includes index.
ISBN 0-19-512364-6 (cloth)
ISBN 0-19-516178-5 (pbk.)
1. Crown-of-thorns starfish.
2. Coral reef ecology.
3. Coral reef ecology—Research.
I. Title.
QL384.A8S27   1999   577.7'89—dc21   98-4634

1 3 5 7 9 8 6 4 2
Printed in the United States of America
on acid-free paper

For my mother,
Susan Atkinson Sapp

CONTENTS

ACKNOWLEDGMENTS

I am grateful for the generosity of many coral reef scientists who lent valuable time in interviews, engaged in correspondence, helped to clarify my thinking, and provided me with a diversity of literature: Charles Birkeland, Roger Bradbury, David Challinor, Richard Chesher, Lu Eldredge, Udo Engelhardt, William Fitt, Peter Glynn, Richard Grigg, Jeremy Jackson, Robert Johannes, Richard Kenchington, Joshua Lederberg, John Lucas, Ernst Mayr, Eric Mills, William Newman, John Ogden, David Pawson, James Porter, Donald Potts, Ira Rubinoff, Stephen Smith, Clive Wilkinson, and Jeremy Woodley. I have also greatly benefited from comments from those who read one or all draft chapters and helped me see ways to make this a better book: Roger Bradbury, Chuck Birkeland, Richard Chesher, Bob Johannes, Jeff Levinton, John Lucas, Tim McClanahan, and John Ogden. I also thank Chuck Birkeland for photographs of the crown-of-thorns in Micronesia. Peter Glynn provided the photographs of coral bleaching and the painted shrimp. John Ogden supplied photographs of *Diadema*. The picture of the triton eating the crown-of-thorns was taken by Johnston Davidson. For the photographs of the crown-of-thorns in outbreak conditions in Australia, I am most grateful to Peter Moran, and the Australian Institute of Marine Science.

I also thank Kirk Jensen and Helen Mules for their editorial expertise. Last and most important, I thank my wife, Carole, and my children, Will and Elliot for their support.

This project was partly supported by a grant from the Social Science and Humanities Research Council of Canada.

INTRODUCTION

There are more things in heaven and earth, Horatio,
Than are dreamt of in your philosophy.

Hamlet, I. v. 174–175

This book is about knowledge and action. It is about the ways in which environmental knowledge is produced and evaluated, and the dilemmas faced by scientists in the midst of uncertainty. What is natural? What is the balance of nature? We address these questions by exploring one of the longest and most poignant environmental controversies in the twentieth century: whether fierce outbreaks of the crown-of-thorns starfish are natural features of coral-reef life, or whether they are caused by human interference. The crown-of-thorns story offers a window from which to examine environmentalism and its relations with marine ecology and governments—from the environmental awakening of the 1960s to the present.

I first learned of the crown-of-thorns in 1989 when visiting the Australian Institute of Marine Science, off the Great Barrier Reef. I had arrived with an interest in symbiosis. Coral reefs were a good place to visit because they are among the most biodiverse communities in the world, aptly compared to rainforests. Cooperative relations among species abound.

Coral-reef communities seem to be so integrated and interdependent that some speak of them as superorganisms. Are such complex tropical systems more balanced and stable than ecosystems with fewer species? Why are coral reefs so rich in species diversity? I had little idea of just how heated these questions had become over the previous two decades. Certainly, fundamental concepts and assumptions at the heart of ecology are at stake. But there is more to this than an academic debate over change in scientific theory. These issues are key to understanding outbreaks of the crown-of-thorns, and in the conservation and management of coral reefs, more generally.

Few marine biologists had ever seen the crown-of-thorns before 1960. These creatures were large, about 60 cm (2 ft) in diameter, covered with sharp poisonous spines, and they were thought to be very rare. They were noticed in plague proportions on one small coral cay, a tourist haven on the Great Barrier Reef. By the end of that decade, they were reported to be present in huge numbers on many of its reefs and to be destroying many other reef communities throughout the Indo-Pacific. They traveled in massive herds of many thousands of individuals, devouring coral and leaving in their wake devastation comparable to a burnt-out rainforest. They mystified biologists. Regarded as one of the strangest ecological phenomena of this century, the crown-of-thorns starfish plagues continued throughout most of the 1970s. They paused for a few years before a second series hit during the 1980s. While infestations remain common on many reefs worldwide, a third major outbreak is making its appearance today on the Great Barrier Reef.

During the late 1960s and 1970s, news of the starfish plagues and their destruction of coral was heralded throughout the world as the kind of disaster predicted by such environmentalists as Rachel Carson and Barry Commoner. They were considered unnatural: the payoff for our careless exploitation of the planet. Many warned that unless something was done to stop their spreading, the population explosions would continue to increase with the most disastrous consequences for coral reefs, many small islands, and their inhabitants throughout the Indo-Pacific.

At the same time, other scientists remained incredulous. Some denied the reality of any starfish population explosions anywhere in the Pacific. Still others admitted their existence, but considered them to be natural, cyclical occurrences, with no long-term deleterious effects. Perhaps they might even be beneficial to coral reefs, and enhance the diversity of life upon them. Therefore, they argued, attempts to control the starfish populations may be irresponsible and result in more harm than good.

Were the outbreaks natural? Or were they human induced? What would be their long-term consequences? What should and could be done to stop them? While these issues captured the attention of many of the world's leading ecologists, political and scientific turmoil persisted. There was much speculation in the press. Special government committees were formed. There were testimonies in the United States Congress, and intense discussions among coral-reef scientists throughout the world. Special crown-of-thorns sessions have been held at international coral-reef symposia over the past three decades.

Discussions of the cause and effects of the outbreaks have involved consideration of virtually every global environmental issue of our time: over-fishing, the heavy use of pesticides in agriculture, atomic testing, the human population explosion and ever-increasing coastal developments, the clearing of tropical forests, as well as the proposal in the 1960s and 1970s to join the Pacific and the Caribbean with a sea-level canal through the isthmus of Panamá. Tropical marine scientists throughout the world are deeply concerned about the destruction of coral reefs. Since the 1980s, the cause and effects of the starfish plagues have been discussed together with another widespread environmental disturbance whose cause remains uncertain: coral bleaching and its association with global warming. Mass mortality of coral due to bleaching has been observed repeatedly and with increased frequency since the 1970s, especially in the eastern Pacific and Caribbean. Whether the increase in coral-reef bleaching is evidence of global warming due to human activities is a question yet to be fully answered.

Environmental disturbances caused by human activity are generally called *anthropogenic*. But we could be more precise. Those changes caused by the overuse of synthetic pesticide sprays, increases in CO_2 from the burning of fossil fuels, CFCs (chlorofluorocarbons), and other greenhouse gases, as well as over-fishing due to "more efficient" fishing technologies, for example, could be called "technogenic." Many environmentalists have challenged the notion of human social progress based on ever-expanding industrial and technological production. Critics often dismiss environmentalists' claims of human-induced global environmental change; they say they are ill founded, and highly exaggerated.[1] The charges and countercharges of environmental and industrial groups are all bids for public support for government action or inaction. Environmental science deals with the relations of knowledge and power, and all agree that understanding the scientific issues in environmental controversies is important for an informed citizenry.

The crown-of-thorns story will take us on expeditions to tropical venues around the world, in and out of marine laboratories, government committees, and technical journals. Much of the early controversy took place in newspapers and magazines. The narratives and images drawn by journalists and scientists provide colorful illustrations of how the crown-of-thorns was portrayed and understood in popular culture. Some government committees, formed to determine the scope and significance of the plagues, were charged with incompetence, deception, and cover-up. Through interviews with leading environmental scientists we will explore their recollections and perceptions of the issues they confront.

Plagues of many kinds have captured public attention over the past few years. A stack of books has reminded us of the Fourth Horseman in the Bible's Book of Revelation.[2] All carry the lesson that nature will avenge itself against those who carelessly abuse it—that pestilence and death will be unwittingly summoned. Outbreaks of new and old diseases defy our former presumption of just a few decades ago—that antibiotics, vaccines, and doctors had saved us from such threats. The dramatic increases in the worldwide movement of people and goods, wars, overpopulation, and pollution have also made the world more vulnerable to ecological disasters. The oceans, not long ago seen as endless resources, can no longer be taken for granted.

Too often, however, those publicizing the reality of new global and human-induced catastrophes appeal to "the balance of nature" as the norm against which human disturbances can be measured. Yet, defining and proving the existence of such a "balance of nature" are the most perplexing issues in community ecology. Although we often hear biologists and environmentalists speak of "the balance of nature," nature itself may be unstable and more unpredictable than is generally thought. Boom and bust, from an influenza epidemic or sudden plagues of locusts to mysterious declines in sought-after fish may actually be the rule in nature. Moreover, ecologists suspect that certain kinds of environmental disturbances may actually be good for ecological communities because they keep species that might dominate from being able to do so, and allow opportunity for many other species to persist. Ecological processes may be operating on temporal and spatial scales that far exceed the scope of most ecological studies. And all this makes studying the causes and effects of large-scale environmental changes, distinguishing between what is natural and what is anthropogenic, and deciding upon what action to take, all the more difficult.

There are still other aspects of complexity to consider. We can approach the crown-of-thorns controversy in much the same way as the ecologist does outbreaks in nature. How do controversies begin? What inflames and sustains them? How are they resolved? Those scientists who studied the crown-of-thorns came from diverse specialties. Different approaches and geographic locations often spawned divergent perceptions and solutions. But when following this story we often encounter bewildering difficulties in distinguishing effects due to the internal processes of science from those due to nonscientific issues.

Critics of environmental activists often insist that we should search for "the scientific truth and nothing but the truth" to resolve environmental controversies. Yet, many environmental problems may require immediate action and cannot wait for the kinds of rigorous demonstrations typically carried out in laboratory science. Many scientists have examined the social and political aspects of the crown-of-thorns controversy at different stages of its development. We will learn of the lessons they drew about public participation in science, and their own behavior in public forums. The tension between environmental "advocacy" on the one hand, and maintaining scientific "objectivity" and professional credibility on the other, is incessant. Untangling the "nonscientific" from the "scientific" in global environmental controversies is, however, often as difficult as separating global anthropogenic change from natural processes. We can identify and talk about them for analytic convenience, but to understand and participate in such environmental issues we need to know how nature, science, and society interact as an integrated whole.

What is Natural?

1

GREEN ISLAND

To find that the little-known inhabitant *Acanthaster planci* had the capacity to destroy its own habitat came as a surprise to many, and there was, in the early stages of destruction, a general reluctance to accept the evidence.

John Barnes, 1966

Acanthaster planci was virtually unknown by coral-reef scientists when Jack Barnes published a short paper in *Australian Natural History* entitled "The Crown of Thorns Starfish as a Destroyer of Coral." [1] Barnes told of how he had come to know it. A medical doctor and naturalist, he had an interest in poisonous marine animals and he was well-known for alerting the public to the dangers of the extremely venomous "sea wasp" or box jellyfish. In the late 1950s, he had become curious about reports of the existence of a large, spiny "stinging" starfish in Queensland waters. But sightings of it were infrequent and he had given up hope of ever seeing one. Even experienced divers considered the seastar a great rarity.

In 1960, special demonstration visits were made to inspect one large specimen that had taken up permanent residence in a patch of low coral on one of the reefs at Green Island, a popular tourist haven on the central part of the Queensland coast. Adult *Acanthaster* grow to about 60 cm in diameter and possess many (9 to 21) arms covered with very sharp spines. Visitors noticed that a few corals had "circular dead patches, bone white and about the size of the star." At first, they were thought to be "long-term resting places" and caused by the "smothering" of the polyps.[2] In the first months of 1962, reports of more and more crown-of-thorns sightings around Green Island reached Barnes. They included stories that its spines had caused human injury and illness to unwary tourists wading on reefs. The symptoms included severe pain for several hours and protracted vomiting every three or four hours for four days.

A low, oval, wooded coral cay, Green Island stands only a meter above high tide. Three kilometers long and nearly 2 kilometers wide, it is made entirely of pulverized coral and sand washed up by wave action from the surrounding reef. Located about 27 kilometers off the coast of the city of Cairns, it is in easy reach of a day boat trip. Visitors came from all over the world to lie on the white sand beaches, swim in the warm green waters, explore the forest and its bird life, or just enjoy a picnic. It was a magnet for naturalists, writers, and photographers.[3] For many, the outstanding attractions were the magnificent coral reefs around the island: the many-colored staghorns, plate corals, anemones, sponges, fans, and giant clams. It was quite easy to explore this coral garden by swimming over it, drifting across it in a glass-bottomed boat, or simply wading over the reef flats at low tide.

Lloyd Grigg managed an underwater observatory on the island. A crocodile hunter, fisherman, diver, and amateur naturalist, Grigg teamed up with Vince Vlassoff, an experienced sea captain and salvage expert, to construct the first effective underwater observatory on the Great Barrier Reef. In 1955, they designed and built a massive, cylindrical 70-ton steel chamber, with thick glass windows. They attached floats to it, tipped it off into the harbor at Cairns, towed it to the end of the jetty at Green Island, sank it to the seafloor, and anchored it down amid a lush forest of magnificent coral and darting shoals of brightly colored fish.[4] It was a very successful tourist attraction for several years. But, in 1962, Grigg told of large herds of crown-of-thorns moving northward on a "front" approaching his underwater observatory. When blemishes appeared near his observatory, he instituted regular day and night inspections. He soon discovered that the coral damage was due to feeding, not "smothering." With constant vigilance he was able to offer partial protection to that small area and make observations on the habits of *Acanthaster.*[5]

By 1963, there was still not much change in the underwater scene. But within the next twelve months the infestation assumed overwhelming proportions, fanning out into coral westward, northward, and eastward of the jetty. Before large herds of *Acanthaster* began to destroy them, these coral gardens could be reached by glass-bottomed boats in all weathers, comfortably and quickly, giving thousands of people an opportunity to see a truly representative sample of the complex reef environment. Boat crews worked strenuously to preserve this area, every week removing by hand as many of the starfish as they could. The starfish tend to wrap themselves around the branches of blue staghorn corals, and boat crews were forced to wear thick gloves and use a steel spike to dislodge them and bury them on the beach. They died quickly out of water—their many feet or arms retract, their sharp spines flatten out, and the body collapses from dehydration; within thirty minutes they are dead.

Despite these efforts, coral destruction accelerated month by month. By the end of 1964, glass-bottomed boats carrying tourists were diverting further eastward ahead of the advancing horde. The "glassies" were taking visitors out to more exposed waters and nearing the practical limits of their travel. Blake Hayles, manager of transport and accommodation facilities on Green Island, decided to concentrate his defense in one selected area, a patch of only 2 acres where he employed a diver to remove the starfish. In the ensuing fifteen months more than 27,000 starfish were taken from that patch. The record for a single day was 373.[6] To many local people, this expensive operation could only delay the inevitable. Although divers managed to save the special coral area visited by tourists, about 80% of the coral on Green Island Reef was destroyed. Such a massive environmental upheaval as the mass destruction of coral by the crown-of-thorns seemed to have no recorded precedent, as far as Barnes could tell.[7] Nor would one expect it. After all, he reasoned, coral occupied a central place in the coral-reef community. It was the main feature around which the diverse other members adapted their behavior.

Coral begins life as a minute unprotected larva swimming in great numbers among the microscopic world of plankton. If the great ocean currents maneuver the larvae to a suitable habitat, they attach to the bottom. Each becomes a polyp—a sea anemone in miniature, complete with a cylindrical body capped by a ring of tiny tentacles and a central mouth. As it continuously divides, each new coral polyp remains attached to the founder by a thin membrane. The polyp secretes calcium carbonate to form a hard white skeleton. Each polyp has its own shallow "hole" in the communal skeleton into which it retreats during the day or when danger threatens. But it cannot withdraw completely. Its soft, colorful tissues remain exposed to the elements and to predators.

Although each polyp stays small, the colony of budding and dividing polyps steadily increases in size and forms massive coral boulders. These coral boulders grow slowly and may require more than a hundred years to reach a diameter of 2 to 3 meters. Coral trees, in which the skeleton becomes a branched or delicately laced structure, grow more rapidly. Through millennia, the steady accumulation of coral skeletons and the skeletons of other reef creatures builds the massive bulk of the reef. They are cemented together by rapidly growing algae, the coralline algae, that secrete a hard, slippery form of calcium carbonate. Together this assemblage of plants and animals creates the form and structure of the coral community.

Coral reefs, nature's most spectacular, exotic, and crowded ecological communities, slowly evolved over a period of almost fifty million years. The largest, the Great Barrier Reef, is about 2000 kilometers (1250 miles) long and composed of about 2000 individual reefs. Several of the Pacific's beautiful coral atolls, small, isolated, ring-shaped islands, are exposed portions of limestone edifices over a mile thick. Their bases rest on the peaks of long-submerged volcanoes.

It was a pretty reasonable assumption that dependent life-forms would be compatible with the continuing existence of coral. So it was no small surprise, Barnes remarked, that "the little-known inhabitant *Acanthaster planci* had the capacity to destroy its own habitat. . . . Unfortunately, the facts are now beyond dispute and after five years of population expansion this spiny seastar . . . is now consuming coral and disturbing the ecology over a wide area." [8]

Barnes described *Acanthaster*'s extremely effective feeding technique. Most large predators find coral unacceptable food. Its thin layer of tissue is so diffused over its irregular limestone skeleton that it cannot be economically harvested. A few fish, crustaceans, and worms nibble at coral colonies, but the crown-of-thorns fed on them exclusively. While most animals must bring their food to their stomach (and thus find it difficult to attack the massive coral boulders), the starfish can bring their stomach to the food. It everts its membranous stomach through its mouth and spreads it over the coral tissue. After it seals off an area equal to its surface coverage, digestive juices pour from its extruded voluminous digestive membrane and liquifies the coral polyps into a greenish slime. When the stomach retracts, only the pure white calcareous (limestone) skeleton remains. The starfish moves on. Algae and other marine growths settle on the porous exposed coral skeleton, the scar turns grey, then green, then dirty yellow-brown. Soon a ragged growth of algae darkens the dead coral. The starfish climbs up from its cover, feeding mainly at night, and Barnes noted that it seemed to have a preference for branching species, especially the large blue staghorns. "Failing these," he wrote, "it selects smaller branching forms, or plate, shelf and boulder corals, in that order."

By 1966, along with the destruction of coral at Green Island, at nearby Arlington Reef more than 9 miles of coral was said to be reduced to debris. The reefs of Michaelmas, Upolu, Clack, Batt, and the Frankland Islands and Port Moresby were reported to have patchy but high concentrations.[10] Barnes thought it reasonable to expect that the population explosion would spread to encompass all the Great Barrier Reef.[11] But biologists were as ignorant of the circumstances that enabled the seastar to escape its normal restraints as they were of the general biology of *Acanthaster*, its natural enemies, or its place in the ecology of reefs.

Coral-reef biology was still in its infancy. Little was known of the requirements of most of the great diversity of species and the relationships necessary for their coexistence on a living reef. Historically, the scientific study of coral reefs began with Charles Darwin's famous voyage on the *Beagle* in the early 1830s.[12] Darwin described corals of many types, obtained some impression of their distribution in depth and the submarine contours of coral reefs, and theorized on coral-reef formation. In the late nineteenth century and early twentieth century, research on coral reefs was based largely on geographical surveys and subsequent museum descriptions of coral skeletons.

The living animal and its ecology were largely overlooked.[13] There were few exceptions. Most scientific knowledge of coral reefs was derived from expeditions. Among the most famous was the Great Barrier Reef Expedition of 1928–29 which established a marine laboratory for thirteen months at Low Island in North Queensland. Led by Sir Maurice Yonge of the University of Glasgow, and sponsored by the Royal Society of London, the records of this historic expedition contained basic data on plankton, the organisms that live on the beds of reefs, and on the metabolism of corals.[14]

After the Second World War, intensive geological and oceanographic studies of atolls in the Pacific took place. They were almost entirely American, driven by military interests and largely concentrated on the Marshall and Caroline Islands. These investigations were intensified before and after the nuclear bomb tests at Bikini and adjacent atolls during the 1950s. There was also a major geological expedition to Kon Tiki Atoll in the Tuamotu Archipelago that provided valuable information about reef formations in the South Pacific and comparisons with Atlantic reefs.[15]

Tropical marine laboratories remained scarce, and huge gaps extend in biologists' understanding of the long-term ecological patterns of coral reefs.[16] So it was not surprising that so little was known of *Acanthaster* and its place in coral-reef ecology. "Until more is learned, " Barnes wrote, "it is difficult to predict where the depredations will end, or what will end them." He was hopeful that this "plague," this "gross imbalance," was self-limiting. In time, he thought, it would exhaust its food supply and then subside. "The reef habitat is enormous, and resilient beyond calculation. Despite the heavy damage in some sections there were survivors, more than adequate to restock the environment as opportunity offers." [17] Other observers were not nearly so optimistic that the plague would be self-limiting and the gross imbalance would correct itself.

In 1963, Barnes contacted the chairman of the Great Barrier Reef Committee, Robert Endean, about human injury attributed to the crown-of-thorns. Formed by the Royal Geographical Society of Australia at the time of the Great Barrier Reef Expedition, the committee was composed of a small international group of naturalists to further scientific research and conservation in the area. It had established a modest marine station on Heron Island in the 1950s. Endean was a marine toxicologist at the University of Queensland in Brisbane. Much of the stunning variety of life on coral reefs was sustained by interactions involving toxic chemicals and venoms as well as attractants. Coral reefs were nature's greatest pharmacological storehouses. In fact, Endean was able to find significant funding for the Heron Island station from the Swiss pharmaceutical giant, Roche. His laboratory in Brisbane also produced a series of headline-grabbing discoveries on the venom of the stonefish and blue-ringed octopus to more far-reaching work on tumor-inhibiting chemicals.

Endean was often in the media and appearing on television, expounding on the wonders of the Great Barrier Reef. He was an outdoorsman and a public figure, with a broad knowledge of the natural history of the reef. However, in the early 1960s, he had never seen the crown-of-thorns and had never heard of people being poisoned by a starfish. After obtaining specimens for investigation, in 1964, he and Barnes published a short note on the venomous spines of the starfish in the *Medical Journal of Australia*, recommending that tourists wear thick shoes when walking on reefs.[18] This marked the beginning of a sharp turn in Endean's career. Research on the crown-of-thorns would lead him from a smooth path of academic distinction into rocky public disputes with government leaders and gov-

ernment agencies—as a champion for the preservation of the Great
Barrier Reef.

Sounding the Alarm

As accounts of *Acanthaster*'s damage reached government authorities involved with the coastal waters, Queensland's Department of Harbours
and Marines commissioned Endean in late 1965 as chief scientific advisor to carry out a scientific investigation. He was to assess the extent of
the infestation, its cause, the risks of the starfish spreading further, and
the means of controlling them. The sum of $A26,400 over two years
was allocated for the study. It was an ambitious project, considering that
the Great Barrier Reef was about 2000 kilometers long and made up of
about 2000 individual reefs that ranged in area from 1 to 250 square kilometers.

One of Endean's students, Robert Pearson, who had just completed a
master's degree, was appointed to do the fieldwork with an assistant.
Their study, begun in April 1966, was extremely productive, providing
data on the extent of infested reefs, feeding rates, food preferences,
growth, movement, fecundity, and breeding seasons of the starfish.[19]
Their survey was designed to check the greatest possible number of reefs
for the presence of starfish populations.[20] They visited sections of
eighty-six reefs (less than 5% of the total number).

No techniques for such a survey had ever been developed, so Endean
devised a procedure. On each reef visited, an observer wearing a
facemask and snorkel or scuba gear would swim slowly in a straight line
for twenty minutes over one or more sections, counting the number of
starfish sighted. Numbers in excess of forty per twenty-minutes' swim
were defined as "infestations." This meant that the population was large
enough to kill 25% of the living coral cover on a reef in one year. This
estimate was based on their studies of the starfish's feeding rates: they
placed individuals in wire cages on reef slopes, fed them branches of living staghorn coral; and measured the area of polyps consumed. They reported "infestations" on twenty-three of the eighty-six reefs visited. The
infestation at Green Island had moved on by 1967, leaving about 20%
of the coral alive. On other reefs, coral mortality ranged from 25 to
95%.[21]

Endean wrote a detailed report and submitted it to the Queensland
government in June 1968. It included a list of recommendations for action.[22] By that time, he had received unpublished reports that the
crown-of-thorns had appeared in large numbers on reefs outside the
waters of the Great Barrier Reef: near New Britain, Samoa, New Cal-

edonia, and Fiji. Starfish were also present in "plague proportions" on reefs near Rabaul.[23]

It was clear to Endean that the plagues were somehow caused by human activities and were a serious a threat to the Great Barrier Reef. Certainly, he considered the possibility that the plagues were natural—that similar plagues had occurred in the past and would occur in the future. But this seemed implausible. There were no previous reports of such infestations, and given the size of the starfish and the damage it caused, it was "inconceivable that such plagues would have gone unrecorded if they are of common occurrence."[24] Moreover, the reefs that were the most heavily infested were those nearest the large centers of human population.[25]

Endean considered three possible means by which human activities may have caused the plagues. All involved the disappearance of predators that would normally keep starfish populations in check. The first was overcollection of the giant triton (*Charonia tritonis*) for commercial trade. To identify predators of the starfish, they had caged it with a wide variety of carnivorous marine animals. The giant triton was the only one that attacked adult *Acanthaster* and Endean suspected that it was possibly the most important factor in keeping down starfish numbers in normal times. There was also evidence that the numbers of triton had decreased as the crown-of-thorns population increased.

The shells of adult giant tritons are extremely large, with an interesting shape, delicately tinted, and patterned. They had been collected for centuries by native people and, since the 1930s, extensively by shell collectors. It was difficult to assess exactly how big the trade in giant triton shells had been because records had not been kept. Tritons were known to have been collected by the crews of luggers, two- or three-masted fishing vessels, engaged in the trochus trade. Trochus resemble very large periwinkles and together with pearl oysters yield a considerable quantity of mother-of-pearl for trade. Trochus fishing in Great Barrier Reef waters began about 1947 and ended about 1960. Available information indicated that the crew of each lugger collected about seventy-five giant tritons per trip, with each trip lasting about seven weeks. With at least a dozen luggers operating at any one time, Endean estimated that about 10,000 giant tritons were collected each year by the crews of trochus vessels. During the 1950s, giant triton shells were sold in souvenir shops in Brisbane and tourist centers in North Queensland. But by the late 1960s shell collectors regarded them as somewhat rare in Barrier Reef waters.[26]

While the overcollection of the giant triton was Endean's favored hypothesis for the plagues, he did consider other possibilities. Tritons preyed on adult and juvenile starfish, but many species of fish would prey on other stages of the starfish's life cycle—on the millions of starfish eggs and the massive numbers of larvae that constitute part of the zooplankton on reefs. Obviously, a balanced ecology would require heavy predation on the egg and larval stages. Perhaps overfishing reduced the populations of such predators. Small fish such as hardheads and sardines, which would prey on starfish eggs, had been caught in large numbers in bait nets. But Endean found no evidence of a decline of bait fish over the previous ten years.[27] Moreover, many different species of potential predators of eggs and larvae still existed; it was difficult to believe that they all had been drastically lowered. Pesticides were also possible suspects. Perhaps large quantities of DDT, dieldren, and endrin had flowed from rivers into the sea, killing predators of the planktonic larvae of the starfish. But again, he could find no evidence of such deaths due to pesticides in the region.[28]

One thing was certain: immediate action was needed.[29] At best, Endean warned, the bulk of damage already caused, might be repaired by coral regeneration within ten to twenty years.[30] He assessed four measures for controlling the starfish plague. (1) Various chemicals (e.g., 5% formalin) could be injected into the body cavities of individual starfish. But this appeared to be impractical. (2) Granular quicklime dropped onto the body surface would also kill the starfish in twenty-four to forty-eight hours. This method had been used to control starfish (*Asterias forbesi*) infesting oyster beds on the east coast of the United States during the 1940s.[31] However, applying it on the Great Barrier Reef would involve considerable expense and effort. Moreover, coral polyps would also be killed by contact with quicklime. He recommended studying an alternative "natural control" method. (3) Large numbers of tritons could be placed on an infested reef. Research would need to determine optimal numbers of tritons and how to breed them in large numbers. In the meantime, he thought there was only one other recourse. (4) Collect the starfish by hand. This would be tedious because the starfish were often intertwined in branching corals; in plague proportions, there might be four or five starfish per square meter of reef. In deeper water around the edge of reefs, collectors would need scuba gear.[32]

Endean's report was submitted to the premier's office in June 1968, but it was not released and published until a year later, and little action was taken. A ban on the collection of triton shells was introduced, and the Queensland government continued to support some monitoring of the situation. Otherwise, the recommendations of the report were not adopted.

THE COMING PLAGUE

Despite the sluggishness of the Queensland government in releasing its report, news of "the Disaster at Green Island" was spread quickly among tropical marine scientists, with the warning that other islands might share its fate.[33] In the American journal, *Earth and Mineral Sciences*, Jon Weber announced in 1969:

> A strange sort of war is raging on Green Island—man against starfish. Unfortunately, the seastar seems to be winning, and the outcome could mean destruction for the entire Great Barrier Reef, as well as many Pacific Islands between the Tropics.[34]

With a background in geochemistry at the University of Toronto, Weber was a specialist in carbonate sediments and sedimentary rocks. He was working out of Pennsylvania State University in the late 1960s, diving on coral reefs on many islands throughout the Pacific. He described *Acanthaster planci* as "an odd animal in many ways, unlike any of its apparent relatives." It was the only genus in the family *Acanthasteridae*, and it was also the only organism known to feed exclusively on coral.[35] Weber told of how for more than two hundred years there had been only occasional reports by naturalists of seeing one or two specimens in widespread locations: East Africa, the Red Sea, Hawaii, the Tuamotu Islands, and the Gulf of California. In 1955, the famous Zoological Museum in Copenhagen had only twenty-one specimens from the entire world,[36] and in only three of the twenty-two collecting sites in the Hawaiian Islands was *Acanthaster* taken.[37] Four specimens had been collected by the *Albatross* expedition to the Philippines, Celebes and Molucca Islands between 1907 and 1910,[38] and only one was found by the 1928–29 Great Barrier Reef Expedition.[39]

At that time, *Acanthaster* was noted only for its large size and unusual appearance. As the naturalist W. K. Fisher wrote in 1925, "It is so unlike any of its apparent relatives that we must regard it, I think, as a holdover from a very ancient fauna—a sort of surviving fossil."[40] But as Weber declared in 1969, now "many naturalists believe that this organism might ultimately destroy the entire Great Barrier Reef, and along with it the unique marine life that it harbors."[41] Although Green Island was devastated, he explained how Arlington Reef, 5 miles wide and 12 miles long, "was reduced to rubble in a few months."[42] Weber himself had discovered concentrations of the crown-of-thorns feeding during the day at Fiji and many other islands throughout the Pacific.

Some scientific reports earlier in the century made no mention of the venomous nature of the starfish, whereas others emphasized how contact with the spines produced extremely painful wounds, redness, swelling, protracted vomiting for several days, numbness, and even paralysis. If these reports were taken at face value, Weber speculated that a mutant strain might exist so that some specimens are venomous and others not.[43] Twice he had experienced the debilitating effects of *Acanthaster's* spines. "The first time, a specimen I was transporting underwater in a nylon mesh "bug bag" was gently brushed against me by an unexpected current. A single spine penetrated heavy gloves and entered the base of my thumbnail. A similar mishap lodged a spine in my knee. In both cases, swelling and pain lasted for well over a week." [44]

That a population explosion had occurred with the most perilous consequences seemed to be as certain to Weber as the venomous nature of the seastar's spines:

> The phenomenon is clearly not localized but widely disseminated. *Acanthaster* has been a rare animal up to now, but if the current population explosion continues, the health and welfare of inhabitants of many Pacific islands will be seriously threatened. Important questions require urgent answers: What caused the current population explosion? What can be done to stop it? [45]

Weber's warning that other South Pacific islands might share the same fate as Green Island was borne out. In 1969, the Queensland government received a request to release Endean's report to the U.S. State Department. Under a technical information treaty with Australia, the Australian government was obligated to share such scientific information if it was requested through official channels. A U.S. Air Force jet flew the report to Washington, D.C., where a small group of scientists and government officials met to discuss plans for an extensive emergency expedition to assess the impact of the crown-of-thorns on coral reefs in Micronesia.

2

GUAM, 1968–1969

There is the possibility that we are witnessing the initial phases of extinction of madreporarian [hard] corals in the Pacific.

Richard Chesher, 1969

The radio crackled to life in Joe Campbell's Marianas Diver Shop. "The water is clear, calm, everything looks O.K." The dive was on! By 10:30 A.M., on Sunday, December 15, 1968, forty-seven divers, U.S. Navy professionals, U.S. Air Force sport divers, and civilian sport divers began the hourlong boat trip along the northwestern shore of Guam to Twin Reefs. It was the site chosen for the first of many battles to open "the new Pacific war."[1] An island-wide "seastar tournament" was organized by the University of Guam's Department of Marine Studies. Navy men submerged with two 90-cubic-foot tanks, the sport divers carried an array of equipment including hamburger tongs, forks, and special spears for picking up the venomous creatures.

After three hours of battle, humans declared an "unqualified victory." The seastars were cleaned out of a beautiful (but now partly dead) reef on the northern border "of the infested zone"—a reef that was slated to become a protected underwater park. Eight hundred and eighty-five seastars were captured, estimated to be enough animals to devour over 80,000 square meters of living coral in a single year. The war had begun. Only a few hundred meters north of the initial battle zone, the divers located a band of starfish kilometers long devouring the living coral from the intertidal zone to a depth limit of coral growth at the rate of 1 kilometer a month. A yearlong control effort began culling starfish from the advancing fronts to prevent their entering the coral gardens to the north and south.[2]

The mobilization of scientists, military personnel, and civilians to wage war on the crown-of-thorns at Guam was due largely to the efforts of a young twenty-nine-year-old marine biologist, Richard Chesher from Scarsdale, New York. Chesher had just arrived in Guam that year and quickly became a leading publicist for the disastrous effects of *Acanthaster* plagues and the need for research and control programs to inhibit their spreading.[3]

The United States had acquired Guam in 1898 as spoils of the Spanish-American War. (It had been a colony of Spain since 1668.) The rest of Micronesia was sold by Spain to Germany, who kept it until the end of the First World War when the League of Nations gave it as a mandate to Japan. After the Second World War, the United States administered the Trust Territory of the Pacific Islands in Micronesia on behalf of the United Nations. It covered a vast area of ocean, just north of the Equator, between Papua New Guinea and Japan, peppered with volcanic islands and coral atolls. Some of them had been bitterly fought over in the Second World War. The Trust Territory included the groups known as the Mariana Islands (running north from Guam to the tropic of Cancer), the Caroline Islands, lying south of the Marianas, and, east of the Caroline Islands, the Marshall Islands, on which nuclear weapons were tested between 1946 and 1958 (Bikini and Eniwetok atolls). The Marshalls later became self-governing, as did the Carolines. Guam had fallen to the Japanese within a few days after Japan attacked Pearl Harbor. It was recaptured by the United States in 1944 and served as an important base for bombing Japan. It was used again as a bombing base during the Vietnam War. The U.S. Navy administered it until 1950, when control was transferred to the Department of the Interior.

The University of Guam was a small college with a two-person biology faculty in the mid 1960s. It was a frontier for the biologists who went there. Typhoons hit regularly beginning in late May and June—sometimes four or five in a year. There were only about two or three stores; if the island was out of sugar, it could be over a month before the next ship came in—before tourism was promoted in the mid-1970s. Between 1967 and 1968, a new science building was erected at the University of Guam. The departments of biology, chemistry, and physics added several faculty members. At that time there was no tenure; scientists arrived on two-year contracts, with an annual expense-paid trip back to their point of departure.[4] Chesher came in 1968, fresh from a postdoctoral year as a National Science Foundation fellow at the Harvard Museum of Comparative Zoology. He remained on Guam for one year.

Chesher had first learned about *Acanthaster* two years earlier when he read accounts of a mysterious population explosion on the Great Barrier Reef. He was then working in the Caribbean as a doctoral student at the Institute of Marine Sciences at the University of Miami. Reports of *Acanthaster* had come from the Red Sea, the Indian Ocean, and the Pacific, but not from the Caribbean. As he later recalled, "The *Acanthaster* problem seemed a remote but interesting curiosity."[5] Moreover, at that time, it seemed to him not to be a problem with any long-term effects. A specialist in the ecology and systematics of echinoderms, the phylum to which starfish belong, he knew that localized population explosions of echinoderms were common and usually short-lived. The oyster starfish, *Asterias forbesi*, was well-known for its periodic destruction of the oyster fisheries on the southern coast of New England during the first half of the century. Its cycles of population explosions and declines had been followed for many years. Based on reports of the oyster industry, newspaper accounts, and testimony of oyster growers, scientists concluded that it had been particularly destructive at intervals of about fourteen years from the 1860s onward.[6]

Chesher's preconceptions and lack of concern about *Acanthaster* fell apart in 1968 as he "hovered over the total wreck of what must have been a truly magnificent reef. . . . I was on the island of Guam, two thousand miles from the Great Barrier Reef. Countless *Acanthaster* paraded below me, leaving dark grey-green ruin where once a fifty-million-year-old ecology had thrived."[7] This was not the kind of localized population explosion reported for starfish affecting oysters in New England. It was obvious to him that the infestations required the attention of many scientists. Capturing their interest became one of his immediate objectives.

Coincidentally, in November 1968, some of the world's leading authorities on the biology of coral reefs gathered in Palau, one of the main islands of the Trust Territory, for a meeting of the International Biological Programme (IBP). Organized in the late 1950s, and launched in 1964 in Paris, the IBP focused on international and interdisciplinary approaches to problems of conservation ecology, biological productivity, and human welfare.[8] Its objective was to ensure the worldwide study of organic production on the land, in the freshwaters, and in the seas, the potential uses of natural resources, and human adaptability to changing conditions.[9] The IBP held international meetings in many countries. It had a special section on marine productivity to improve understanding of the basic ecological mechanisms that control the abundance and distribution of marine organisms in inshore areas where the effects of human activities were so great.

The coordinator of the Marine Productivity section was Sir Maurice Yonge, leader of the historic 1928–29 Great Barrier Reef Expedition. He developed a theme on coral-reef research and conservation.[10] Other leaders in coral-reef science at the IBP meeting on Palau included Thomas Goreau, from the University of the West Indies, Jamaica, and Siro Kawaguti, from the Seto Marine Laboratory, Japan. They met for a few days in Palau, and then the biologists at Guam invited them to have two sessions at the University of Guam. At Chesher's request, they agreed to take a look at the *Acanthaster* infestation.

Both Yonge and Goreau had recently seen *Acanthaster*. Yonge had observed populations in Borneo.[11] Goreau had first observed them in 1962 when he had participated in an expedition in the Red Sea funded by the U.S. Office of Naval Research, and the Department of the Navy and Department of Zoology, Tel-Aviv University. On his return from the expedition, he published a paper on the starfish's extremely effective feeding techniques, while recommending further studies of its ecology.[12] He noted how it seemed to feed mainly at night, hid in dark crevices by day where it could remain for twelve to seventy-two hours before coming out to feed again, and how it could move great distances over bare sands with a high crawling speed of up to 10 meters (33 feet) per hour.

The possible impact of *Acanthaster* on the growth and development of coral reefs did not escape Goreau. He noted that in the Red Sea, around Entedebir, Um Aaback, and Sula Bay, there were no large coral reefs, despite ideal conditions for corals. Yet, there were fossil remnants of enormous reefs that had thrived in the region not long before. He suggested that *Acanthaster* might be responsible for large-scale devastation by seeking out and destroying enough of the fast-growing young coral colonies to keep the rate of framework construction down to a level at which no net reef accretion could occur. Thus, Goreau called for its further study, "in view of the strong probability that this species may, under certain conditions, be an important factor limiting the growth and development of coral reefs." [13]

Chesher got some of the results he had hoped for. On returning to Palau, Yonge, Goreau, Kawaguti, and others resolved, at the International Biological Programme Conference, to recommend immediate research into the outbreaks.[14] Senator Richard Taitano also made a series of site inspection dives with Chesher. On the senator's recommendation the government of Guam commissioned Chesher to carry out an emergency, six-month research program to study the "invasion" and attempt to control its spread.[15] With an initial budget of $US15,000, Guam government funds were used to employ a team of divers to try to halt the advance of a "front" down the east coast of the island. They found long bands sometimes broken up into groups that moved as "amorphous herds" of up to two hundred individuals. "One large adult could in a single night clean off a coral head that required fifty years to grow." Tagging showed individual movements of up to 250 meters (825 feet) per week.[16]

Despite their declared victory in protecting one reef, by March 1969, 90% of the reefs were dead along 38 kilometers (24 miles) of the coastline of Guam. Starfish were "killing the reef" at a rate of about one half-mile per month.[17] Chesher and his colleagues experimented with various methods of killing the invaders. Finding them to be difficult to dislodge from branching corals, they decided that injecting them with some form of poison would be the most effective technique. Formalin and ammonium hydroxide proved efficient killers, and when dispersed in the ocean after the death of the starfish, appeared to have little effect on other marine life. Chesher developed a device, like a hypodermic syringe, with a long needle that automatically refills itself on the end of a spear. By operating the handle, as with a bicycle pump, a lethal dose of a chemical could be injected into a starfish in a few seconds. A scuba diver could kill a hundred or more in an hour. Even a snorkeler could dispose of up to ten with a single breath.[18]

The more Chesher studied *Acanthaster*, the more concerned he became. The effects of the large populations at Guam were the same as those observed at Green Island in Australia. When the coral died, a community of algae smothered the skeletons and changed the entire complexion of the reef from a world of pastel colors to a drab, inert graveyard. The myriad reef creatures that had been evolving for millions of years, adapting to the living coral reef, suddenly faced an alien environment. The alteration of the total environment by a change in color and texture shattered the animal associations. Fish and mobile invertebrates vacated the dead reef. Algae smothered many of the smaller filter feeders. The caves and crevices of the reef became clogged and overgrown as algae bloomed everywhere. *Acanthaster* killed the coral and the subsequent imbalance destroyed most of the associated fauna.[19]

In Chesher's view, *Acanthaster* plagues were dangerous to the whole reef community, and on more remote islands, that community included humans. Many islanders obtained all of their protein from the sea. Even in technologically advanced areas where outside protein sources were available, fishing and the coral gardens were valuable tourist resources. Coral was also, of course, instrumental in maintaining reefs, which protect the coastline from erosion during storms.[20] He thought that the very land, the atolls, upon which islanders stood might be in danger. Atolls are small ring-shaped coral islands enclosing a central lagoon. Hundreds of them dot the South Pacific, consisting of reefs several thousand meters across.

During his voyage on the *Beagle*, Darwin had proposed that many atolls were formed on ancient volcanic cones that had subsided, with the rate of growth of the coral matching the rate at which the inactive volcanoes subsided into the seafloor. His explanation was confirmed one hundred and twenty years later when scientists working for the U.S. Geological Survey, who were conducting extensive drilling programs on coral reefs, hit volcanic rock hundreds of meters down.[21] Over millions of years, the coral formed limestone caps up to a mile thick. Occasionally, the corals added to the island more slowly than the rate of subsidence and these atolls submerged, now existing as guyots.[22] Many atolls are only about 1 or 2 meters above sea level and are quite small. Chesher considered that erosion of even a small portion of shoreline by storm waves would be a serious threat.

Knowledge of coral-reef dynamics was negligible. It was impossible to give an accurate assessment of the consequences of the massive coral predation by *Acanthaster*. Indeed, Chesher was surprised and distressed when he began to realize just "how little was actually known about the delicate ecology that was in danger." [23] In the spring of 1969, the Guam government agreed to support another moderate program to kill starfish over the next fiscal year, but there were no more funds for research. Worse still, Chesher could find no other marine scientists actively working on the problem anywhere in the Pacific. He contacted Robert Endean at the University of Queensland in Brisbane, only to find that research there had come to a halt and Endean's report and recommendations for control measures had still not been acted upon—or even released by the Queensland government. Endean informed him that the invasion was still going strong on the Great Barrier Reef. Chesher also received reports from amateur divers and shell collectors of infestations in numerous areas including Borneo, New Guinea, Fiji Islands, Truk, Palau, Yap, Rota, Saipan, Wake, and Johnston Island. But it seemed to him that in some of the reports, normal populations of starfish were mistaken for infestations. A standardized search was needed. Scientists and legislators in the United States had to be advised about the need for action.[24]

A CALL FOR ACTION

In May 1969, Chesher sent a paper to the most influential scientific journal in the United States, *Science.* It was quickly published two months later under the title: "Destruction of Pacific Corals by the Sea Star *Acanthaster planci.*"[25] Chesher explained how *Acanthaster* had a voracious appetite for living coral but had been regarded as a great rarity until about 1963, when huge swarms were destroying large tracts of coral on the Great Barrier Reef. He emphasized how Thomas Goreau had singled out the crown-of-thorns as an explanation for the impoverished coral growth observed in the Red Sea. He pointed to devastating outbreaks on Guam and Palau as well as reports from amateur divers about devastation on many other Pacific islands. All the available information indicated "that recent population explosions of *A. planci* are occurring almost simultaneously in widely separated areas of the Indo-Pacific Ocean and that these are not short-term population fluctuations of the type reported for numerous other marine invertebrates."[26]

Chesher called for better estimates of the extent of the infestations and the severity of the damage, and research into the cause of the population explosions. He was skeptical that predation by the giant triton would normally control the starfish populations. When he had penned two tritons in with an adult *Acanthaster,* they often ate only half the starfish; the remainder escaped and lived to regenerate lost parts. And one triton attacked only at a rate of one seastar every six days. "Even if the triton were abundant," Chesher argued, " it is doubtful that it could control *A. planci.*" Moreover, large populations of the seastars were found in areas seldom visited by shell collectors and where tritons were common, such as parts of Palau and Rota.[27]

Endean was looking at the wrong stage in the starfish life cycle, in Chesher's opinion. The cause of the population explosions had to be sought in the predation of the starfish's larvae. Based on the life histories of other starfish, he estimated that one female would produce between one and twenty-four million eggs. Endean had considered fish that preyed on the larvae. But Chesher pointed to coral itself as their main predator. The larvae swim about in the open sea for about twenty days before settling down in shallow water and changing into small starfish. In most Pacific areas, shallow waters suitable for seastar settlement are blanketed with the tentacles, traps, and snares of an endless variety of creatures that filter floating organisms from the seawater. Chesher suspected that these filter feeders were a major regulator in the checks and balances of the reef community. Almost all members of the reef environment, including corals, send millions of offspring into the ocean currents each year. Only a tiny fraction of their progeny survive the rigors of pelagic life, and the filter feeders of the reef kill most of these survivors when they attempt to settle out of the plankton. Corals, sponges, clams, and sea squirts are only a few organisms that eat not only their own species, but the young of their reef mates. When tiny seastar larvae land on living coral, the tentacled coral polyps snare them and then devour them.

Corals, along with other filter-feeding animals of the reef, were unquestionably the most important predators of starfish, in Chesher's view. When corals were eaten by starfish or destroyed by any other means, their death would provide ripe conditions for larval settlement. Once an area of reef was killed, a settling ground would be provided for any swarm of larvae that would attempt to settle there. The starfish would then move out into neighboring areas, killing more coral and consequently extending the areas in which more larvae could settle. Subsequently, the starfish would produce a new large crop of larvae that, if the currents were right, could flood the dead and dying reefs with still more seastars. The end result would be a population explosion. The fantastic numbers of larvae produced by the steadily rising population might then inundate neighboring coral reefs with so many young seastars that even a normal, living reef might become overwhelmed.

When Chesher first arrived on Guam, significant coastal developments provided the very conditions he prescribed for the population explosion. Flat coral reefs were blasted open to make boat channels and harbors dredged to make them deeper. The infestations in Guam, Rota, and Johnston Island were first observed near blasting and dredging activities. This was no mere coincidence. Destruction of reefs by blasting, dredging, and other human activities, he argued, had "provided fresh surfaces, free of filter feeders, for settlement of the larvae. In such areas, original populations of several hundred animals, concentrated together, might provide the necessary seed population for an infestation. Such dead coral areas must probably be freshly provided during time of larval settlement (December and January in Guam)." [28]

The strategy for controlling infestations would follow directly from this cause. Adult starfish had to be prevented from infesting new coral areas. Long-term control might be possible by monitoring areas subject to blasting or dredging during periods of larval settlement. Seed populations had to be eliminated before larval settlement the following year. If infestations were detected at an early stage when seed populations were localized, this would be simplified. When found at a later stage, Chesher suggested, the adults might be held in contained zones where they would be left to starve to death after a period of six months. Where such containment zones could not be set up, sections of reefs could be protected by local extermination of the starfish. He told of how, at Guam, advancing fronts were staved off by weekly inspections of a 2-km coastline. Divers, towed behind a boat, killed migrating sea stars with an injection gun containing full-strength formalin. [29]

Chesher did not shy away from prophesying disaster for the people of small isles and atolls of Oceania if something were not done: "Most inhabitants of Oceania derive almost all their protein from marine resources, and destruction of living reefs results in the destruction of fisheries. Eventually, loss of living corals would allow severe land erosion by storm waves." [30] It was also possible, he warned, that the wholesale destruction of coral would continue to the point where the coral fauna could not recover. *Acanthaster* predation might lead to the extinction of hard corals throughout the Pacific.

PROJECT STELLEROID

The doomsday predictions in Chesher's *Science* article echoed throughout the tropical marine science community. But he did not stop there. In May 1969, at the end of the emergency six-month program funded by the government of Guam, and less than a year after he first observed the outbreak, Westinghouse Ocean Research Laboratory (WORL) offered him a research position and asked if he could come to California for an interview. He was to carry out a study of the environmental impact of a desalination plant in Key West, Florida. During the interview process Chesher explained that he could not leave Guam until there was a satisfactory solution to the problem of the *Acanthaster* plagues. He knew that his position as associate professor at Guam could do little, yet he felt it was irresponsible to walk away from the problem, and a waste of time to simply continue swimming about pruning starfish off the reefs. Endean had written him telling about how his report had been buried by officialdom. As Chesher recalled, when he received the phone call from WORL the whole strategy emerged in his mind all at once. He went to interview "with WORL with a single goal—to awaken the scientific community to the issue of the impact of man on coral reefs and ignite research using the crown-of-thorns as a focal point." [31]

Newly established in 1966 in the Sorrento Valley, 15 miles north of downtown San Diego, WORL was part of the Westinghouse Laboratories in Pittsburgh, the central research and development organization of the Westinghouse Electric Corporation. Along with a broad "in-house program," Westinghouse undertook considerable research and development for the government.[32] WORL was created to carry out an oceanographic program with emphasis on basic research problems that would supplement or aid the corporation's rapidly expanding engineering and operational involvement in ocean-related businesses.

WORL's proximity to the United States' oldest and best established oceanographic institution, the Scripps Institution of Oceanography of the University of California, San Diego, in La Jolla,[33] coupled with the spectrum of oceanographic conditions found off the coast of California, provided an ideal environment for its oceanic program. Though only three years old, it had acquired experience carrying out such studies as measuring waves and ocean acoustical characteristics, as well as monitoring deep currents at 2000 meters.[34] Most pertinent to the crown-of-thorns problem, it had been engaged in monitoring biological communities and the physical environment off the coast of southern California to determine natural and man-induced changes in the nearshore ecology.[35]

During Chesher's interview, the whole staff of WORL became intrigued with the starfish story. The response was a fine example of how a corporation can take decisive and immediate action. The director of WORL suggested that they fly to Pittsburgh to meet with the vice president of Westinghouse Research Laboratories. From there they flew to Washington, D.C., to meet with the vice president of Government Affairs. Within hours of arriving in Washington, they were having talks with high-level officials in the Department of the Interior because of the possible economic implications for the Micronesian Trust Territory. Tourism was considered to offer more for the future economic development of the U.S. Trust Territory than any other single prospect. With improved air transportation, the tropical Pacific was an area growing in popularity for tourists and vacationers. Viewing the reefs by snorkelling, scuba diving, and other means was one of the main attractions. Coral reefs and clear waters could attract substantial revenue.

Discussions were also held with representatives of the Office of the Science Advisor to the President, and the Marine Sciences Council. In less than a week WORL scientists consulted with other tropical marine specialists, including Thomas Goreau, Porter Kier, chair of the Department of Paleobiology, U.S. National Museum, Smithsonian Institution, Washington, D.C., and J. E. Stein, zoologist at Boston University and research director of the New England Aquarium. They wrote a proposal and got tentative approval from the Department of the Interior to proceed with their plan.

They intended to conduct an expedition led by an international body of marine scientists who would immediately converge on the U.S. Trust Territory to ascertain population levels of *Acanthaster,* determine the extent of coral damage, and gather data on possible causes and controls of the infestations. Because of the rapidity with which starfish could "kill irreplaceable coral reefs," the survey had to take place that same year. Scientists could only participate during summer months when they were free of university obligations. September in the mid-Pacific is typhoon season. All these factors determined that the survey had to take place between July 1 and August 31.

The Westinghouse proposal to the Department of the Interior laid out the stakes in a way that mirrored those in Chesher's *Science* article:

> Islands within the U.S. Trusts are severely threatened by the starfish *Acanthaster planci.* There is a widespread population explosion of these poisonous starfish which feed on living coral.[36]
>
> Destruction of the living coral reefs would be an unparalleled economic disaster for the smaller isles and atolls of Oceania. Loss of protective living corals will permit severe land erosion by storm waves. Also, most inhabitants of Oceania derive the major part of their protein from marine sources; consequently, destruction of the living reefs will result in the elimination of a vital food source." [37]

The proposal was successful. On June 6, 1969, the U.S. Department of the Interior allotted $225,000 for the survey. Chesher was appointed Chief Scientist of the project working for WORL. The U.S. Navy, the U.S. Coast Guard, and the National Science Foundation further supported the project. The Office of Naval Research supplied some scientific participants, but mainly coordinated logistic support. The Smithsonian Institution and the National Science Foundation organized the scientific personnel. Scientists at the University of Hawaii put together a companion effort to survey the Hawaiian and Marshall islands.

June was a flurry of planning and organizing. The single largest logistic problem Chesher faced was transportation between islands. Some islands, such as Kapingamarangi, were hundreds of miles from the nearest airstrip. At the last minute, the U.S. Navy made two seaplanes available to supplement their own air-sea rescue vehicles; final plans were worked out just before participants arrived in Guam on July 1. About fifty participants were to be organized into ten survey teams, debriefed, and within a week or so travel to investigate sixteen widely scattered remote islands covering an area larger than that of the continental United States.[38]

Logistics was only one problem. Recruiting qualified biologists who had experience in coral-reef research was another. Individuals came from Australia, Puerto Rico, Venezuela, and various institutions scattered throughout the United States. WORL sent six participants: specialists in photography, physical oceanography, logistics, finance, and environmental management. Six members of the University of Hawaii participated. Only fifteen participants had Ph.D.s in ecology or marine biology, and about nine of them specialized in some aspect of tropical marine science. About an equal number of coral-reef specialists were graduate students. The largest group was centered around Thomas Goreau at Discovery Bay, Jamaica. A glance at the research environment at Discovery Bay will betray the nascent state of coral-reef biology in the late 1960s and of some of the problems facing its development.

DISCOVERY BAY

Goreau is a legendary figure in the history of coral-reef science. He began college as a medical student before completing his Ph.D. in ecology in 1956 at Yale under the direction of another legendary ecologist, G. Evelyn Hutchinson.[39] Goreau first visited Discovery Bay in 1951 and established a small marine station there in 1965; by then, he had enormous prestige as a leading coral physiologist and ecologist. He taught physiology to medical students at the University of the West Indies in Kingston. He was also an excellent diver, known to go down 100 meters to collect corals on a single tank of air.

Goreau's experimental research in the early 1960s on the growth of coral tissues and coral skeletons was highly celebrated.[40] Central to it was the question of the association of the animal polyp and its symbiotic algae. Maurice Yonge had shown how tiny unicellular algae penetrate the coral tissue at birth and live there in a symbiotic relationship.[41] The algae, containing chlorophyll, function as chemical factories converting sunlight, carbon dioxide, and organic nutrients (metabolic wastes from the coral) into oxygen and carbohydrates. Embedded in coral tissue, they are protected from predators and gain supplies needed for protein synthesis: nitrogen and phosphates that are so scarce in the nutrient-poor tropical waters. Corals also benefit from the algae by obtaining oxygen and carbohydrates and having their metabolic wastes recycled back into fuel within their bodies.

All reef organisms benefit from this delicate symbiosis, for, without it, coral reefs and coral islands would never have been formed. Indeed, Goreau demonstrated that the algae were essential for rapid skeleton formation, playing an intimate role in calcium metabolism. He and his wife Nora, a biochemist, showed that corals with symbiotic algae grew faster in sunlight than those deprived (by being kept in the dark) of their algae. In a series of brilliant experiments using radioactive calcium-45 as a tracer, they were able to measure skeletal growth by examining how much calcium was taken from seawater — ultimately to make the calcium carbonate skeletons.[42] But Goreau was an eclectic scientist. He could work happily in zoology, botany, ecology, and biochemistry, which was just what was needed in coral-reef studies at the time. During the late 1960s, a group of eager young graduate students and researchers gathered around him, not just those from the University of the West Indies, but from Yale and elsewhere.

Jeremy Jackson was one of several students who joined the Westinghouse expedition at Goreau's invitation. At that time, he was a Ph.D. student in the biology-geology program at Yale, where he worked with the paleontologist Donald Rhoades. Before completing his thesis, he was advised that if he wished to work in the tropics he should go to work with Goreau.[43] In the summers of 1968 and 1969, Jackson began his career-long interest in the study of brachiopods, and what he calls cryptic communities in coral reefs—the little organisms that encrust underneath corals. Brachiopods were the dominant marine forms of Paleozoic (570–248 M.Y.A.) and Mesozoic (248–65 M.Y.A.) times, and a few species survive.

There was plenty of research about understanding ancient and modern reefs at Discovery Bay in the late 1960s. Willard Hartman, from Yale, discovered certain kinds of sponges, indicating that paleozoic coral reefs were built up by sponges as much as by corals. Judy Lang, another Yale student, discovered interspecific aggression by slow-growing corals that attack faster-growing species, showing that the abundance of a coral species was not determined by sheer growth rate. Henry Rieswig found that sponges were nutritionally unlike anything anyone had ever seen before. David Barnes, a physicist from

Newcastle-upon-Tyne, worked on coral skeleton growth and, together with his colleague Janet Lough, later gave coral-reef scientists a new understanding of annual growth bands in corals.[44] Jackson recalls how Goreau's father, Fritz, the great science photographer for *Life* magazine, would come in periodically and they would spend a few days helping him set up a photo.[45]

Yet, at that time, the Discovery Bay Marine Laboratory was little more than a shack on the beach on the east side of the bay. In 1969, construction began for a new marine laboratory with a main building and workshops. But the construction and operation of the new laboratory met with serious difficulties. Indeed, the problems encountered illustrate some of the issues limiting the growth of coral-reef marine science in developing countries.

Jeremy Woodley was director of the new laboratory between 1975 and 1992. As he recalls, because of Goreau's reputation two things happened in 1968. "One was the university was able to raise funding in Britain (from the Wolffson Foundation) to build Tom a real lab on the western side of the bay where it is today." [46] However, due to a devaluation in the British pound at the time of building, some plans had to be abandoned (including a proper room for a library). The land was donated by the Kaiser Bauxite Company. The second thing that happened was that the Marine Science Center at the State University of New York (SUNY) at Stony Brook, Long Island, came "head-hunting for Goreau." At that time, SUNY was establishing a new Marine Science Research Center and a Department of Ecology and Evolution. The outcome of the negotiations was a compromise. They got half of Goreau, and he did not leave Jamaica. He was jointly appointed as Professor of Marine Sciences at the University of the West Indies (UWI), and Professor of Biological Science at SUNY. His salary was paid by both universities, and SUNY also agreed to help fund the operation of the new marine laboratory.

The partnership was a farsighted agreement. It assured the new facility in Discovery Bay of top-class professional direction, with an American institution and UWI bringing their expertise and students. It also ensured that UWI would be helped in meeting the running costs that were always a problem plaguing the development of third-world institutions. As Woodley remarked, "Agencies are happy if they have funds to build something that they can put a brass plate on and stick their name on it, but they walk away. And it is yours to find the money to pay the staff, the electricity bills and the rest of it." [47]

The new laboratory was opened in March 1970, under the awkward although effective symbiotic title: "State University of New York–University of the West Indies Marine Laboratory." Goreau was its director. Unfortunately, a month later he died of stomach cancer, at the age of forty-seven. That tragedy completely changed perceptions of the marine laboratory. Over the next five years, SUNY hardly used the facilities. By the mid-1970s, it decided to opt out of the arrangement completely. It was a severe blow to the University of the West Indies, which found itself with a big institution, but no funds. To remain operative, the marine laboratory had to meet most of its own costs. Under Woodley's direction, and with the flying start provided by Goreau, over the next years the still-modest laboratory was able to build up a clientele of various American universities who brought classes and visiting researchers who paid fees to use the facilities.

That many young coral-reef scientists were willing to put aside their own research in the spring of 1969 to travel to Micronesia said much about the concern over *Acanthaster* and the authority wielded by Goreau. After a month back in New Haven, Goreau telephoned Jackson asking if he would like to join the expedition. Jackson had known something of *Acanthaster* only because Goreau had given him a copy of his reprints when he first visited Discovery Bay. Goreau not only believed *Acanthaster* was a serious problem, he also had enormous regard for Chesher. As Jackson recalls, "Tom had a lot of respect for Rick's ability, his work on echinoids was excellent; he was a superb photographer, and a skilled writer." [48] Judy Lang, Eileen Graham, and one of Goreau's sons, Peter, an underwater photographer, went in one group with Thomas Goreau to Saipan. Jackson went with a group led by Woodley.

Woodley had arrived at the University of the West Indies in 1966 shortly after completing his Ph.D. at Oxford on the functional anatomy, ecology, and behavior of brittle stars. He had met Chesher as a graduate student when he visited the United States in 1963. They had worked alongside one another at the marine laboratory of the University of Miami. Later, when Woodley heard about Chesher's activities in Guam and that he was trying to assemble a team, he asked to join. But, as Woodley later remarked, assembling "thirty or forty diving biologists who knew what coral reefs looked like . . . was not that easy! Today you'd get 3000–4000." [49] He himself had little experience with coral reefs at that time. As an undergraduate student and member of the Oxford Exploration Club, in 1959, he organized a six-month, twelve-member expedition of naturalists to what was then British Guyana. Led by evolutionist Arthur Cain, the group studied the adaptive radiation of frogs, birds, insects, plants, and their diversity and the way they lived.

That expedition gave Woodley a taste for working in the tropics. After arriving in Kingston, he familiarized himself with the natural surroundings and later began to visit Discovery Bay. As he put it, "I was impressed by coral reefs and bemused by them." [50] He was appointed team leader on the grounds that he had a Ph.D. The others in his group were graduate students: Jackson, Barnes, Morgan Wells, Yushia Neumann, a student from Israel, and Asterio Takesy, a native of Truk and student at the University of Guam. They went to Truk.

MICRONESIAN SURVEY

A three to four-day orientation period in Guam was arranged. Participants surveyed the Guam reefs to familiarize themselves with search techniques and, with advice from the Australian survey, established a survey procedure. They then departed for the islands carrying equipment for life support and diving in remote conditions: inflatable boats with outboard motors, scuba tanks and a compressor, charts and data on their destinations, special logbooks, and countless smaller items including a shark gun. Each team was to:

- Estimate the size and area distributions of existing populations of *A. planci*
- Examine the effects of these animals on the coral
- Seek information relating to possible causes of infestations
- Note unusual features of the marine environment
- Observe feeding habits and behavior of *A. planci* [51]

They estimated population size by recording the number of animals seen per twenty-minute tow or swim. A "normal" population was considered to have fewer than twenty specimens per twenty minutes of search.[52] Divers counted starfish, and an observer in the boat recorded the total number seen at the end of the tow. "We called it 'trolling' because there were so many sharks," Jackson recalled. "I took off one flipper and put one foot on the cowling of the outboard motor, and held onto the line of the zodiac, and could get to that boat in a microsecond." [53] The islands of Truk were arranged in a donut shape with several dead volcanic islands in the center surrounded by atolls. Americans had sunk every single Japanese supply ship in that part of the Pacific in the lagoon. It was all there, and Jackson saw thousands of sharks. But what he saw in regard to *Acanthaster* was devastation. "The herds started off around high islands in the middle, and were working their way around in two directions. They were obviously going to meet and starve to death." [54]

By August 15, the field studies were completed and scientists debriefed. Before they returned home via Guam, they prepared a report and sat for hours discussing their findings. The debriefing sessions were recorded and transcribed. "Suddenly it was September, and three man-years of raw data" engulfed Chesher's office. By mid-October he had completed his analysis of the data and written a lengthy report. In assessing the results, he placed considerable reliance on the subjective impressions related during debriefing sessions.[55] Comments on the conditions around Guam were illustrative: "far worse than I suspected" . . . "depressing," "fantastic," or some variation of these.[56] The vast majority of the data was easy to interpret: "Either the starfish were quite scarce and the reefs normal and healthy or the starfish were present in vast herds which were doing obvious and wholesale damage to the living coral." [57]

The data was discouraging. Ten islands had sufficiently high populations to be considered infested: Saipan, Tinian, Truk, Pohnpei, Rota, Palau, Ant, Guam, Majuro, and Arno. Johnston Island, Kapingamarangi, Nukuoro, and Pingelap were questionable areas, with high population levels of starfish that needed to be examined at a later time. Yap, Ifalik, Woleai, Lamotrek, Kwajalein, Hawaii, Mokil, Midway, Kauai, Oahu, Maui, and the French frigate shoals were found to have normal populations.[58]

Chesher emphasized that the analysis and conclusions reached in this report were his own as chief scientist of the project.[59] But he was careful to draw on as much consensus as possible; and consensus seemed to run deep: "Everyone who observed the destruction felt the situation was clearly an important phenomenon representing an extreme unbalance and an unnatural ecological condition." [60] Yet, the data proved only that massive herds of *Acanthaster* were devouring large tracts of Pacific coral reefs. The cause or causes of the infestations remained uncertain. Chesher weighed the strengths and weaknesses of various theories.[61] He began by considering whether the plague was a natural part of long-term reef ecology. If this were the case, he reasoned, "the periodicity of the cycle must exceed 100 years and possibly 1000 years." [62] He supported this statement with three kinds of evidence.

First, there was the lack of scientific reports of such massive infestations. Chesher insisted that the absence of such evidence had to be taken as evidence of the absence of past infestations. The seastar's large size, sharp spines, shallow-water habitat, high population densities, and profound effects on coral reefs made it improbable that past infestations would have gone unrecorded. The startling contrast of the pure white, freshly killed coral against the living reef would hardly have escaped attention. During their surveys some participants interviewed Micronesian islanders to learn if such infestations were part of the folklore of people who had been fishing on the reefs for more than 2000 years. Islanders were aware of the starfish and had a name for it in their particular dialect, but they seemed to know it only as an uncommon inhabitant that one must avoid stepping on.[63]

One also had to consider the size of the coral killed and the time it took for it to grow. Goreau estimated that some of the corals killed represented several hundred, possibly more than 1000 years of continuous development. Allowing time for regrowth after a previous hypothetical explosion, Chesher argued that a period exceeding 200 years would be the minimum for a cyclic recurrence of *Acanthaster* devastation.[64] It was "possible, although rather pointless," he argued, "to construct a theory that accepts a 200-year cycle for an invertebrate that would recur on an interocean scale." [65] One also had to consider that almost all the infestations began near human populations.[66]

Most researchers suspected that the infestations were caused by some change in the environment that led to the release of predation pressure on some stage in the seastar's life cycle.[67] There were a number of ways this could have happened. Perhaps pollution had caused a decrease in the populations of predators that would feed on *Acanthaster* eggs and planktonic larvae. Concentrations of organochlorines and other human-made pollutants had been increasing in the marine environment during the previous few decades. This was an area in dire need of study.[68]

In the meantime, Chesher found further evidence to support his hypothesis that blasting and dredging or continuous mechanical damage to coral reefs would provide settling areas for *Acanthaster* larvae, leading to a sudden population explosion.[69] Blasting, to open up harbors and channels, had been common on several islands that survey teams reported as infested: Pohnpei, Rota, and Nukuoro, as well as Guam. Blasting with dynamite had been a familiar method of fishing on Truk since the Japanese occupation; and Truk had one of the oldest infestations. During the spawning season of 1968 a channel was blasted through the reef near Cocos Island at the southern tip of Guam (in an area outside the existing infestation). When divers inspected the area after the completion of the channel there were no *Acanthaster*. When they inspected the site five months later, they found numerous specimens between 6 and 8 centimeters in diameter.[70]

There were also problems with Chesher's theory. Destruction of sections of coral reefs by humans had been common since before the Second World War. And numerous blasts occurred on the reefs during the war without population explosions. Storm damage from typhoons also produces fresh coral surfaces, and they occurred annually. But in Chesher's view, these kinds of disturbances would not allow enough time for the kind of settlement and recruitment necessary for a population explosion. Bombings or storms would be much less likely to provide a suitable surface than regular, methodical efforts to obtain fish or open a passage or harbor using explosives. He thought it would be a rare occurrence for a typhoon to strike a particular reef and cause significant coral damage during the very time *Acanthaster* larvae were abundant. Moreover, seed populations of starfish seemed to begin on the leeward side of an island, normally protected unless a typhoon hit the island directly.[71]

There was still another weakness in Chesher's theory. It seemed to apply to Guam, Rota, Ponape, Truk, and Nukuoro, but not Australia.[72] To reconcile this apparent anomaly, he reconsidered the overcollection of tritons. Tritons ate slowly and were not always fatal to adult seastars, but Endean suggested that their predation on young starfish would be much more effective. Whether this actually occurred was not known. "In theory," Chesher admitted, "depletion of triton stocks might be responsible for the population explosion." [73]

Chesher's report for Westinghouse considered one more possibility. Perhaps radiation fallout from atomic tests in the South Pacific had caused a mutation to occur in the starfish that improved its ability to survive. For example, *Acanthaster* larvae (like many marine invertebrates) possessed the ability to seek out adults to settle near. Perhaps this ability had not been present or well developed, and appeared through mutation. But this seemed very unlikely. For, if a mutation were the cause of the infestation, an advancing wave of infestation should occur following the several currents that would carry the new strain of starfish. Yet there was no apparent orderly pattern to the outbreaks. The random distribution of infestations pointed to local disturbances.[74]

Collection of tritons, local destruction of reefs, and pollution were all "plausible explanations." One simply needed more research. In the meantime, Chesher recommended that a control program be initiated immediately. It would include an active eradication effort on infested reefs of economic or scientific value along with an educational program to inform islanders how they could contribute to research and control efforts.[75] He insisted that it was "not necessary to identify the causes before implementing limited controls to halt infestations." [76] Nonetheless, some observers who did not participate in the survey advised, "Let

Nature take its course," or "It's a natural phenomenon and doesn't need control" and "Let's study it for awhile." [77] But Chesher insisted that control programs should be established whether the infestations were natural or not. In his view, nature's course would be no more desirable in the case of *Acanthaster* infestations than in an uncontrolled forest fire. Controls were necessary to protect the welfare of local inhabitants.[78] After all, he argued, locust plagues and other biological epidemics were produced naturally and no one doubted the value of control over those natural catastrophes.

The risks of not acting were high. Like Endean, Chesher suspected that it would take at least twenty-five years for any noticeable recovery in coral cover to occur. Moreover, "recovery" assumed there would be no further reinfestation of a damaged reef.[79] The possibility remained that coral fauna would simply not recover. To appreciate this possibility one has to consider the special nature of coral-reef communities. While temperate ecological communities usually contain large numbers of a few species, the inverse was true of tropical reef communities—the ocean equivalent of tropical rain forests. (Today, the Great Barrier Reef is recognized as home to some 1500 species of fish, 4000 mollusks, and 400 corals.[80] A single reef may contain as many as 3000 species.)[81] Although reefs contain a wide variety of coral species, numbers of any one species on a particular reef are usually low. Therefore, if destruction of coral on an island occurred on a broad enough scale, Chesher warned, it could be an exceedingly long time before such species were able to reestablish their populations (if they ever could).[82] Reestablishment of an area by corals to a climax population of a lush coral reef was thought to be a long and complex process. But no studies had ever been carried out over long periods. There was little evidence to go on.[83]

One issue was certain: destruction of coral disturbed the entire reef community. Field teams confirmed that large food and game fish were almost totally absent on "dead reefs," and the majority of brightly colored "tropical" fish were missing from the algae-covered reefs.[84] There was no immediate danger of atolls "washing away," but Chesher warned that repeated breakage of dead coral by storm waves could result in wave erosion of portions of the shoreline.[85] Thus, he recommended a three-pronged strategy:

1. Train Micronesian divers to use scuba, and build, maintain, and use formalin guns to protect valuable reefs.[86] Toxic fences, vibrating fences, and electrical barriers also warrant evaluation as a means of containing *Acanthaster*. Such control measures would be

temporary—until more detailed studies were completed and long-term biological controls established.[87]

2. Administer an education program to alert islanders to the problem and how they could help prevent loss of living coral on a civic basis. All participants in the survey agreed that an educational movie on the *Acanthaster* problem could be introduced into the high school system in Micronesia.[88]

3. Increase scientific investigations of the biology of *Acanthaster*, its predators, and the dynamics of reef degeneration and regeneration. Studies of the seastar's behavior, physiology, predators, parasites, and diseases would be essential to the development of long-term biological controls. Studies of the biology of corals themselves were needed. Field teams were as surprised as Chesher about just how little was known about the dynamics of coral reefs.[89]

CONCERTED ACTION

Chesher submitted a preliminary report to a review panel organized by the Department of the Interior, which met October 9–10, 1969, at the Scripps Institution of Oceanography, La Jolla. The panel included Chesher, Goreau, Robert Jones, director of the University of Guam's Starfish Control Project, two members of Scripps (its chairman, Marston Sargent, and the well-known tropical marine biologist, William Newman) as well as two officials from the National Science Foundation. Considering the threat to fisheries, and especially to the growing tourist industry, they called for prompt control measures "wherever severe damage with consequent economic loss" was occurring or seemed imminent.[90] They also recommended that funds be made available for an intensive research program on *Acanthaster*. Long-term monitoring was required to characterize normal conditions and to establish a baseline from which to measure damage and reef recovery. They suggested that a crown-of-thorns program manager be established at one of the three major funding agencies: the National Science Foundation, the Office of Naval Research, or the National Institutes of Health, and that a steering committee be formed to coordinate activities in the Pacific.[91] They estimated that "an adequate research, monitoring and control program in areas of highest interest to the United States" would cost about $830,000 per year, and take several years to complete.[92]

The following month, on November 19, 1969, four members of the

U.S. Senate, Hiram L. Fong and Daniel K. Inouye from Hawaii, Henry Jackson from Washington, and Gordon Allott from Colorado, introduced a special bill into the Senate for $4.5 million for a research and control program for the crown-of-thorns problem.[93] At Senate hearings on the Fong Bill, on March 18, 1970, Robert Jones testified:

> Today, while this hearing goes on . . . thousands of these starfish are devouring coral species on the island of Guam and our neighbor islands of the Trust Territory. The people of Guam are delighted to see this bill presented and to observe its progress through the orderly and precise channels of democracy. We look forward with great hope to its eventual passage and with anticipation for the help it will bring us.[94]

As senior scientist with WORL, Chesher also testified the same day:

> The facts are undisputed: coral reefs are dying. They are dying rapidly and in many island areas. Coral reefs are valuable. We must find out if we are responsible for this ecological upset. We must determine what will happen if the reefs die. While doing this, we must protect the more valuable coral reefs with a well-planned control program.[95]

Senator Fong, representatives of the Smithsonian Institution, the Department of the Interior, and Guam's representative in Washington also testified. Both the Department of the Interior and the Smithsonian Institution strongly favored enactment of the legislation. The Department of the Navy, on behalf of the Department of Defense, had no objections. The Committee of the Whole House on the State of the Union was unanimous in urging its prompt passage. The bill was subsequently referred to the Committee on Merchant Marine and Fisheries, which recommended it.[96] Passed by the Senate and the House of Representatives, it was sent to President Nixon for signing.

3

THE WAR OF THE WORLDS

H. G. Wells could scarcely have bettered the sense of helplessness and chilling menace induced by the huge and venomous starfish, *Acanthaster planci*, now teeming in the waters of the Pacific. This evil-looking creature has razor-sharp spines that can easily pierce a leather glove, and can cause extreme pain, vomiting and even paralysis. For totally obscure reasons, *A. planci* is undergoing massive population explosions almost simultaneously in widely separated areas of the Indo-Pacific Ocean. The immediate result is the complete disappearance of vast tracts of coral, for which the starfish have a voracious appetite. Already, some 140 miles of the Great Barrier Reef north of Australia—more than a quarter of the total reef—has disappeared. Now the creature is on the move throughout the Pacific. According to marine biologists in the area, if the devastation continues unchecked, the atolls and reefs, the islands they protect, and the marine life they shelter, could be destroyed.

Bernard Dixon, *New Scientist*, 1969.

The Martian invasion of Earth described by H. G. Wells in *The War of the Worlds* (1898) is a prototype for science fiction dealing with the threat of invasion.[1] Like all good science-fiction writers, Wells had a solid scientific education and kept the company of, and even collaborated with, some of the leading scientists of his day. And like all good science fiction, *The War of the Worlds* incorporated contemporary science and social issues. The plot of the book emerged out of discussions with his brother Frank about their doubts that civilization had the ability to meet crises bravely and intelligently—this book was to deal with that problem.[2] Written at a time when the British empire was threatened, the opponents, largely Bismarck's Germany, lurked just across the channel. Their invasion of London was a more likely possibility in the minds of readers than the possibility of inhabitants on Mars. But what the Martians looked like, how they behaved, and whether the citizens of Earth would conquer them, appealed to popular imagination.

As the book opens, Mars was undergoing environmental upheaval, suffering from global cooling. The Martians flee to establish a new home. Landing near London, the insect-like beings, equipped with a heat ray, repel Earth men. Later, the reader observes the cryptic Martians close up as they peer out from under a destroyed building. Like *Acanthaster* hiding in a coral crevice, they are equally as strange. They consist mostly of head, with no visible nostrils, a beak, and two large eyes. They have sixteen tentacles near the mouth, used as hands, but they are mostly brain and lungs evolved for the thinner Martian atmosphere. On Earth, they eat by injecting blood from animals directly into their veins; on Mars their diet is suspected to be a human-like beef cattle.

The book's narrator ridicules the English for discussing methods for warding off the invaders while doing nothing. He likens them to the dodo lording over its nest discussing the arrival of the sailors "in want of food." "We will peck them to death tomorrow, my dear." While the Martians, buried under the top layers of the earth, begin to build a flying machine to conquer more areas, a British artilleryman offers some comments on survival: the need to preserve the species until we can learn enough about the invader, the necessity to have controlled breeding and education to reach the level necessary for survival. Alas, he was all talk and no action.

The flight of the English before the monsters was the beginning of the rout of civilization, "the massacre of mankind." As historian David C. Smith comments, readers of the book do not think this is excessive, as they learn of the breakdown of morals, ethics, and the brutalization of the frightened English.[3] The narrator receives various responses from those fleeing the Martians. A curate comments that their plight was simply that foretold in the Bible's *Book of Revelation*.

Diligent, twenty-four-hour-a-day workers, intelligent, and technologically advanced, the Martians were no match for the lowliest Earth creatures. Luckily for humans, bacteria attacked and killed them all. Human evolution on Earth by natural selection had left humans, but not Martians, resistant to bacteria. But Wells' moral was plain: nature would not always be so favorable to humans. The end of the world, caused by evolution, or extraterrestrial events, though not necessarily by Martians, was inevitable. As Londoners pick up the pieces and restore their world, the narrator preaches his lesson: "We have learned now that we cannot regard this planet as being fenced in and a secure abiding-place for Man; we can never anticipate the unseen good or evil that may come upon us suddenly out of space." Wells emphasized that it was necessary to support scientific research, "the conception of the commonweal of mankind," and that our success would come from anticipation, education, and the concerted efforts of our species, not by frantic efforts.

The plagues of the crown-of-thorns, the unthinkable disaster, and the historical context that made them so inevitable were in many ways similar. The threat of nuclear bombs in the Cold War had already affected perceptions of the future by showing the real threat of global catastrophe. The leading British ecologist Charles Elton made the analogy with "population explosions" explicit in the introduction to his well-known text, *The Ecology of Invasions by Animals and Plants*, of 1957:

> Nowadays we live in a very explosive world, and while we may not know where or when the next outburst will be, we might hope to find ways of stopping it or at any rate damping down its force. It is not just nuclear bombs and wars that threaten us, though these rank very high on the list at the moment; there are other sorts of explosions . . . ecological explosions. An ecological explosion means the enormous increase in numbers of some kind of living organism—it may be an infectious virus like influenza, or bacterium like bubonic plague, or a fungus like that of the potato disease, a green plant like the prickly pear, or an animal like the grey squirrel.[4]

Reports of the crown-of-thorns population explosions also occurred at a time when environmental matters were first becoming major international political issues of considerable emotional impact following the writings of such authors as Rachel Carson and Barry Commoner, as well as Paul Ehrlich's warning in *The Population Bomb* about the outbreak of our own species.[5] Carson, the fountainhead of the environmental movement, had already alerted the world to the hazards of using pesticides. Her 1962 book, *Silent Spring*, also warned of other problems resulting from upsetting the balance of nature: outbreaks of imported plants such as prickly pear in Australia, plagues of spruce bud-worms in Canada, and the decline in salmon stocks. But because of the destructive use of insecticides, she prophesied that, "With the passage of time we may expect progressively more serious outbreaks of insects, both disease-carrying and crop-destroying species, in excess of anything we have ever known."[6] A silent spring bereft of bird songs may surely come, she warned. Would her predictions also apply to the already silent, but beautiful and elaborate coral gardens of the world? Chesher followed in Carson's footsteps in heralding *Acanthaster* infestations, and calling for action.

WAR IN OCEANIA

In *Skin Diver*, under the title "Divers Wage War on the Killer Star," Chesher told of "a war, in Oceania, against a by-product of man's haphazard treatment of his world. Against an animal, an enigma; once only a rare, curious-looking member of coral communities, now a biological nightmare whose population has exploded beyond that delicate point of equilibrium that balances construction and destruction."[7] But whether "guilty or not," he argued, "man must try to save the coral which is so vital to the way of life of many Pacific islanders." If the reefs ever grew back, he warned, it might take several hundred years before they reached the present stage of development. He called on divers who enjoyed the Pacific coral reefs to join together, as had divers of Guam, in efforts to locate the menace and prevent further expansions. Under a photograph of the enemy, Chesher wrote:

> Pacific divers: If this seastar is becoming very common in your area, please write to Dr. Richard Chesher, University of Guam, P.O. Box EK, Agana, Guam 96910. We are interested in knowing just where these areas are and how extensive this problem is becoming. Please include as much information as possible on when the spiny seastar started to become numerous and if shelling or dredging is common or was common a year or so before the infestation began.
>
> Do not attempt to kill the animals by cutting them up. This may not always be fatal to the seastars. The best way to kill them is to pick them up and get them out of the water. The spines are venomous and quite painful (but not deadly) so handle them with a knife or spear and be careful to dispose where they cannot injure someone.[8]

In the third volume of *Oceans*, Chesher published *"Acanthaster: Killer of the Reef,"* announcing that war had officially been declared by the American government with the introduction of the special Senate bill for a research and control program on the crown-of-thorns:

> On November 19, 1969, three members of the United States Senate proposed an all-out war on a bizarre red and green, 16-armed predator. Perhaps through man's rape of the global environment, the crown-of-thorns starfish has begun devastation of some of the most valuable real estate in the world—the coral reefs of the South Pacific.[9]

Again he explained how a skin diver's spear was an effective way of removing the starfish when they plague underwater parks. But they could be killed much more effectively by injecting them with formaldehyde contained in a syringe that automatically refills itself on the end of a spear.[10]

At the University of Guam, Lu Eldredge, chair of the Department of Marine Studies, began the *Acanthaster Newsletter* to keep researchers up to date on events while news in the area funneled through a new temporary marine laboratory, established with local starfish monies. In February 1970, the ground was broken for the University Marine Laboratory. By July, the Marine Resources Division of the Trust Territory had begun training Micronesians in scuba diving and organizing starfish control teams in the Mariana, Palau, Truk, Pohnpei, and Marshall island districts. Control efforts continued under the supervision of the Territorial Division of Fish and Wildlife. Nearly 25,500 starfish were killed between October and July 1970. Sixteen thousand were removed from the water; the others were killed with formalin guns.[11]

Control teams were also established in the district centers of Micronesia. More than 15,000 starfish had been collected from Saipan. From Rarotonga, Cook Islands, came notice of an alarming increase in *Acanthaster* in the lagoons and reefs of Manihiki, Penrhyn, Pukapuka, and Palmerston atolls, and near the harbor at Rarotonga. It was feared that the mother-of-pearl was affected by the starfish, because there seemed to be fewer young oyster shells than usual. The fisheries officer in Apia, Western Samoa, informed Eldredge that the south coast of Upolu was infested, and a bounty equivalent to 8 American cents was established. By March 1970, some 13,873 starfish were collected along a 5-mile reef. *Acanthaster* also appeared on the south coast of Savaii. It turned out that the palolo worm population decreases when the starfish appears—palolo is such a choice delicacy that the villagers were more than willing to kill the starfish.[12]

There was news in spring 1970 that a small shrimp, *Hymenocera elegans*, popularly known as the harlequin shrimp or painted shrimp, feeds on the crown-of-thorns.[13] This was discovered by coral-reef scientists at a scientific meeting in Tanzania. *Time* magazine covered the story.[14] Wolfgang Wickler, a behavioral biologist at the Max Planck Institute, was discussing animals that form lasting bonds with their mates —such as jackals, gibbons, geese, and the painted shrimp. Almost in passing, he mentioned that the shrimp feeds on starfish, including the crown-of-thorns. At the request of biologists at the Smithsonian Institution, Wickler later set up a demonstration of an encounter between a large starfish more than a foot across and a pair of the 2-inch-long shrimp.[15] Oblivious to the starfish's poisonous spines, the shrimps quickly lifted one of its arms and began tickling the tiny tubular feet of their prey. Instantly, the starfish retracted them, effectively immobilizing itself. After a few minutes of joint effort, the shrimps succeeded in toppling the crown-of-thorns onto its back. They punctured its tissue with their sharp pincers and proceeded to tear out chucks of flesh. After a full day's feeding, the starfish was reduced to a pile of jellied debris and spines.

Several films also depicted starfish studies, and television news magazine shows covered the plagues. The Westinghouse Ocean Research Laboratory produced a twelve-minute film entitled "Search for the Killer Starfish," which portrayed the Micronesian survey in the summer of 1969 and discussed some possible causes. It was filmed mainly on Guam, with some scenery shots of Yap and Palau, and showed the University of Guam survey headquarters. In May 1970, an NBC special, "The Great Barrier Reef," was aired. It included the shrimp segment. Another film, lasting about twenty minutes, was made in Hawaii by the Department of Land and Natural Resources and a local television station. It showed a large population of *Acanthaster* in the Molokai area, and Division of Fish and Game control measures. On February 18, 1970, the Australian Broadcasting Commission devoted its thirty-minute program "Four Corners" to the "Menace Down Under." Robert Endean was interviewed.[16] These were merely some highlights of the publicity the plagues attracted. Chesher had designed, staged, and directed the Westinghouse survey to focus global scientific interest on man's impact on coral-reef ecosystems. He encouraged the public media and invited *National Geographic* on the expedition. The war in Oceania, along with the testimonies of scientists who participated in the Department of the Interior's expedition, carried the crown-of-thorns into living rooms. Magazines and newspapers, from the *Micronesian Reporter* to the *New York Times*, heralded news of the destruction, risks of further disaster, the heroic efforts of divers to prevent the starfish from spreading, and the need for further coordinated action on behalf of the commonweal.

THE ALIEN PREDATOR

Many had been well prepared by the writings of environmentalists on the dangers of unrestrained industrial and technological development, and the unchecked growth in human population. Whether the plagues were due to radiation fallout from atomic testing in the South Pacific, overfishing, shell collecting, dredging and blasting activities following the Second World War, or excessive use of DDT and other insecticides, one thing seemed certain. Humans had unwittingly upset the delicate balance of nature through their reckless exploitation of the environment.

Journalists preyed on the seastar's common name, "the crown-of-thorns," its gigantic size, its alien form, and its venomous spines: "Few creatures are more aptly named," wrote a reporter for *Time* magazine. "The crown-of-thorns, large reddish brown sea dweller, has as many as 21 arms, all covered with venomous spines that can temporarily paralyze a swimmer and provoke fits of vomiting." [17] The religious aspect of the "crown-of-thorns" and the suffering of the unsuspecting and innocent Pacific islanders were highlighted in the *Micronesian Reporter*:

> For nearly two thousand years, the expression "crown-of-thorns" has been linked vividly with agony, pain and sorrow . . . somehow, during the past few years, someone or something has disturbed the reef ecology and man now is scrambling to regain and retain the balance of life in coral areas . . . and the term "crown-of-thorns starfish" also is beginning to mean agony and sorrow for Pacific islanders. [18]

The name "crown-of-thorns" seemed appropriate if worn by the island people of the South Pacific, or by the coral reefs themselves, but a reporter for *The Illustrated London News* argued that it was inappropriate if worn by the starfish:

> "Crown-of-thorns starfish": the name has romantic overtones to anyone familiar with the story of Christ and fittingly suggests a tragedy. Yet, the name is hardly appropriate. Moreover, it seems to be of recent vintage. . . . The spiky starfish hit the minor headlines in 1968–69, but pests, like celebrities, tend to emerge into the limelight and then disappear from the news. [19]

Few reporters took the starfish plagues so lightly. The effects, often compared to plagues of locusts, seemed to be of biblical proportions. [20] All the deadly animals of the sea that instilled fear into the hearts of humans paled in significance. "Sharks? Sea snakes? Sea wasps? These are the creatures that flash into many minds when someone mentions killers on the Great Barrier Reef. But now a different kind of killer has arrived on the scene, one whose onslaught may affect the very existence of the entire reef." [21] "It may well turn out," wrote one commentator, "that modern science may be up against one of its most formidable challenges—that of trying to save the Great Barrier Reef—a feat that would be unparalleled in our time." [22]

That the geological structure of reefs, and of Pacific islands themselves were at risk was featured by many reporters who covered the survey in Micronesia. On June 12, 1969, the *Washington Post* reported:

"A medium-sized atoll could be wiped out in five years," says Dr. R. D. Gaul, manager of the Westinghouse Ocean Research Laboratories in San Diego.

"It's a real disaster, " says Dr. Raymond Fosberg of the Smithsonian Institution. "We're not only going to lose the coral, but everything that depends on them—including hundreds of species of fish that provide protein food for Pacific islanders.

In addition, the coral reefs that surround many Pacific islands protect them from eroding under the pressure of wind and waves.

A 15-month study of sea life on the Great Barrier Reef in 1928 turned up only one specimen. Now 127 of the starfish were found on a five square-mile sample of the reef.

One U.S. Interior Department official worries that the starfish may be carried into the Caribbean's coral beds. "Something has happened to the world ecology that has upset the delicate balance whereby this very efficient coral predator was kept in check," says Dr. Gaul.[23]

The threat to the coral gardens of the Caribbean, and comparisons with other doomsday scenarios, brought the issue closer to the backyards of American readers. The next month under the headline "Battle of the Coral Sea," *Newsweek* compared the threat of *Acanthaster* to prophesies four months earlier when storm warnings were up for the entire state of California. It was predicted that California would be separated from the mainland by convulsive earthquakes and drift out to sea, thus making Boise, Idaho, the country's chief Pacific seaport. "California lives. But now the island-state of Hawaii—and other Pacific islands—may be in danger from natural forces of a different sort—ravenously hungry starfish." [24]

Porter Kier, chair of the Paleobiology Department of the Smithsonian Institution, was quoted in *Newsweek*: "People will starve. . . . It could affect all the marine environments of the world." Kier attributed the outbreaks to "too much DDT, too much dynamite, and too much residual radiation from past atomic tests." That the Queensland government had not released Endean's report further fueled intrigue:

> The Australian Government conducted a study of the crown-of-thorns . . . but for reasons not explained the findings have been kept secret. And so late this month Kier and some 60 other scientists will join Chesher on Guam for their own survey of the life style of the starfish in order to find methods to control its population explosion. But the destruction it has caused is irreversible. Says one scientist: "Nobody has ever seen a destroyed reef come to life again." [25]

PROMETHEUS UNBOUND

For many readers, the real alien in this war was technology itself. The lopsided development of the physical sciences and technology was an egregious illustration of "the age-old heresy of man's worship of himself." Historians of technology in the late 1960s appealed to the Judaic story of the fall in the Garden of Eden, and to the Greek legends of the demigod Prometheus and of Daedelus to warn of the perils of exploiting knowledge purely in the worship of material achievement and unreserved faith "in the religion of science and technology." Yet, as David Landes wrote in *The Unbound Prometheus* in 1969, the point that humans would be punished for their presumption was only part of the truth of such ancient tales: knowledge was retained.[26] Adam and Eve lost Paradise for eating the fruit of the Tree of Knowledge; but they maintained the knowledge. Prometheus was punished for stealing fire from Olympus; Zeus sent Pandora with her box of evils to compensate for the advantages of fire, but Zeus never took back the fire. Daedelus, who made wings for himself and his son Icarus, lost his son. But he was the founder of a school of craftspeople and sculptors, and passed much of his cunning on to posterity.

If there was no turning back the expansion of science and technology, it was clear to many that the development of the physical sciences and technology was not balanced by ecological knowledge of our natural environment and social knowledge of ourselves. In the life sciences, genetics occupied the central position. At its core was the discovery in 1953 of the molecular structure of the "secret of life," the spiral staircase of DNA. In the 1960s, problems in biology that could be explained in terms of chemistry and physics were often hailed as the only ones that ranked as having any serious scientific merit. This was a time when such popular writers as Isaac Asimov told the "intelligent man" that biology is a system proceeding from chemistry to the associated subjects of genetics to neurophysiology. All else was considered "stamp collecting." Asimov's 1960 best-selling two-volume treatise, *The Intelligent Man's Guide to Science*, lacks even mention of the word ecology.[27] Sociologist of science Dereck de Solla Price boldly asserted in *Science* that he agreed "firmly with Asimov about what is central in science and what is not and will defend him to the death against traditionalists who may deplore his not starting with 'heat, light and sound' or his giving short shrift to natural history."[28]

In the early 1960s, when the International Biology Program began to organize itself around ecosystem approaches to ecology and conservation, it was looked upon with derision and ridiculed as merely representing "a gimmick to keep up with the physicists." As the IBP chief organizer C. H. Waddington recalled, "The general idea among biologists was that ecology dealt with a blow-by-blow account of the day in the life a cockroach, woodlouse or sparrow; and the notion that it could study such questions as what does the ecosystem do with the incident solar energy tended to be greeted with blank stares." [29] Molecular biologists, Waddington asserted, could care less about the IBP, but were apprehensive that it might take away some of their public funds.[30] From the 1960s onward, molecular genetics spawned research into genetic engineering, gene therapy, DNA-based pharmaceutical research, and more recently the program to locate and map all the genes in the DNA of a prototypical human being. The latter aim was often compared with placing a man on the moon.[31]

On July 20, 1969, when the Earth (not Mars) was considered to be cooling and "with mother nature on the run," the lunar module *Eagle*, carrying Neil Armstrong and Edwin Aldrin, landed on the moon in the area named by astronomers centuries earlier "The Sea of Tranquility." Millions of people watched on television as Armstrong became the first person to set foot on the moon. The Sea of Tranquility was purely fictitious and misnamed by the medieval astronomers who imagined it. But down on Earth, it was not difficult to imagine a disaster unparalleled in the history of mankind. The real seas of the South Pacific were far from serene, equally as unknown, and mysterious.

The space race in the Cold War, like the Manhattan Project (to build an atomic bomb before Nazi Germany) which preceded it, gave physics great momentum. There was nothing in ecology to compare. The cultural distance between ecology on the one hand, and physics and molecular biology on the other, was almost as dramatic as that between the space-age world and that of the South Pacific islanders themselves. This striking juxtaposition, this war of the worlds, was captured as an historical moment frozen in the *New York Times* on July 21, 1969: "'The Eagle has landed.'—Neil A. Armstrong." The quotation was surrounded by a long article under the headline: "Scientists Say Coral-Eating Starfish Peril Pacific Islands":

> The voracious 16-limbed sea animal is said to threaten the food supply, and even the physical existence, of many other islands and atolls . . .
>
> "If the starfish population explosion continues unchecked, the result could be a disaster unparalleled in the history of mankind," Dr. Richard Chesher . . . said. One scientist called it "a real war between the species."
>
> On these sun-drenched, unspoiled coral islands the brown-skinned

Micronesians live much as they did centuries ago, with a minimum of clothing and a simple diet consisting mostly of fish and coconut.

Dr. Chesher . . . and other marine scientists here have warned that the destruction of protective reefs by the starfish will expose the low, sandy islands of the Pacific to irresistible battering by storm-driven waves and the normal surf, which are now restrained by the natural coral barriers.

The result, Dr. Chesher said, will be the gradual disappearance of such islands, considered among the world's great beauty spots, as livable places.

But before the Pacific peoples suffer the physical erasure of their home islands from the surface of the sea, he said, they will face starvation through the loss of food resources from their reefs.

Chesher believes that the starfish can be brought under control. He has proposed an intensive educational program throughout the affected islands. "We must make the starfish menace part of the local folklore," he declared. "Killing starfish must become a ritual practice with the natives." [32]

It was not just atolls that were in danger. Geoffrey Harrison, chief fisheries official of Queensland, reportedly told a group of American scientists on Guam that "if the starfish attack the outer-side of the reef, all the ports of northern Queensland state will be doomed." [33]

The Centre for the Biology of Natural Systems at Washington University, St. Louis, was concerned with such environmental hazards as the relation of air pollution to rates of cancer, and the increase in Asiatic flu epidemics. [34] It perceived the outbreaks of *Acanthaster* as representing only one more warning that unless the United States learned quickly to apply its scientific knowledge to problems of the environment, it courted "inevitable disaster." This would occur because our ability to acquire and apply technical knowledge far outstripped our biological ability to adapt to the resulting changes and relate properly to our environment and ourselves. If left unresolved, the consequences would be our own extinction: "Man may blow himself out of the universal scheme by mishandling nuclear fission; he may drown himself in his own protoplasm; or he may choke his species to death in the effluent of his affluence." [35]

"South Pacific Nightmare," *The Economist* called it, on November 22, 1969, explaining how the coral reefs of the Pacific were "crumbling, and the economies of whole regions could crumble with them." [36] The starfish "plague" was "caused either by the products of the chemical industry or the apocalyptic weapons made by physicists." In the long term, "the coral atolls may be actually washed away." More immediately, the livelihood of large numbers of people was already disappearing along with the coral. In the wake of the devouring starfish, the fish that live among the coral move out into deeper, rougher waters. Because islanders were equipped only with primitive outrigger canoes, they were unable to follow them. *The Economist* explained that "in less than three years,"

> many of the best fishing grounds have already been wiped out, especially in some of the worst hit areas around the American trust islands of Tinian and Saipan. Reports are coming in every week of new areas hit by the plague. They include Malaysia, Midway, Hawaii, some of the 800 islands of Fiji, and the Solomon Islands; Fiji and the Solomons are supposed to be under British protection. So far, the islanders do not seem to realize what has hit them. They have not asked for the sort of help they ought to claim, like loans for more seaworthy fishing boats.[37]

Wells's doomsday scenario was caused and ultimately settled by events outside the control of humans. But that seemed to be a remote possibility in the case of the crown-of-thorns. *The Economist* spoke for many when it asserted that the most credible theories lay the blame on some sort of human interference with the environment: "High concentrations of DDT in the ocean could have killed off some of the starfish's predators, particularly at the larva stage. Just conceivably, fall-out from some of the Pacific nuclear tests could have triggered a change in the life-cycle of the starfish." [38] Emergency funds and concerted international research on the cause of the infestations were needed immediately to control the starfish plague:

> The Americans have allocated $250,000 for research. Britain, France and Holland—all of whom have responsibilities among the Pacific islands—have done nothing. This is not to their credit. Although the full damage from the plague may not show itself for perhaps another 20 years, it is in the next five years that the damage will be done. And the time to make a major international effort to counter it is now.[39]

In July 1969, Chesher was confident that they were winning the war against the starfish in Micronesia with "his injector gun." [40] But others soon became less optimistic. In March 1970, the *Washington Post* explained how Guam was losing the fight against the starfish invasions, even though bounty hunters on Guam were paid 25 cents for each starfish killed.[41] At the University of Guam, Jones compared attempts to eradicate the crown-of-thorns on an individual basis as "a flyswatter technique." [42]

HYPERBOLE

Although journalists and science popularizers acknowledged the scientific uncertainty over the exact (anthropogenic) cause, they often failed to recognize the same for the effects. Speculations about the risk of erosion of the physical structure of reefs, and of shorelines and islands being washed away, often became hardened facts when set in newspaper ink. Thus, under the headline "Reef Victim of a Blunder" the *Miami News* explained how the damage to the Great Barrier Reef threatened some six hundred islands and islets resting atop it as well as numerous towns and ports that rely on the reef as a breakwater against the Coral Sea and the Pacific.[43]

Journalists also often went beyond scientists' own hyperbole and public vilification of the starfish. With the ignorance about corals and coral reefs, and in the midst of this hyperbolic frenzy, the crown-of-thorns was frequently portrayed as actually eating away at the physical structure of the reefs. "Starfish Eat Reefs Because Man Has Upset Nature, Study Says" was the headline in the *Washington Post* on December 10, 1969, reporting on the senators' bill to allocate $4.5 million over the next five years to control the starfish.[44] Under the "B" science-fiction title: "Giant Starfish Devour Pacific Reefs," the *Miami Herald* reported how on Guam and Rota, "200,000 of the ugly monsters are chewing away at the coral."[45] *The Economist* spoke, in the same terms, of the effects on the tourist economy around the Great Barrier Reef, which attracted 400,000 visitors a year. It was crumbling as was the reef, because starfish were "eating their way through it at the rate of about 40 miles a year." Unless some spectacular way was found to eliminate the starfish, "the greater part of the Barrier Reef may be eaten away within a decade. That would expose the Queensland coast to the full force of erosion by the ocean."[46]

The amount of coral reported to be eaten or destroyed, especially on the Great Barrier Reef, seemed boundless. Endean and Pearson had visited only 5% of the reefs of the Great Barrier Reef, and then only sections of them, to make extrapolations that about one-quarter of those reefs visited were infested. Yet the editor of the *New Scientist*, Bernard Dixon, wrote in 1969 that "more than a quarter of the [Great Barrier] reef has disappeared."[47] When reporting on the bill to provide $4.5 million for a five-year control-and-study program, the magazine *Ocean Industry* pointed to another "50 square miles of coral confirmed dead on the Great Barrier Reef raising the total mortality to *1000 square miles*." Typically, the article carried the subheading: "Very existence of many islands threatened. Cause of infestation still not determined." [48]

In Australia, "the crown-of-thorns plague" was broadcast widely and repeatedly in newspapers and magazines. In the *Courier Mail* (Brisbane) in July 1969, Endean was quoted as saying: "The fact is the Reef could go . . . If the reef goes the Queensland coast will be exposed to the Pacific Ocean." Two months later, in *The Australian,* under the heading "Reef Collapses After Starfish Attack," Endean "said the whole reef could die off and be broken up and dissolved in the ocean within 50 years." Such reports resulted in a widely held public fear, exemplified in a poignant letter in *The Australian* in 1969 that asked: "How will we be able to look our children in the eye when they ask, 'What was the Great Barrier Reef?'" [49] In Australia, the anxiety over the cause and effects of the infestations, and calls for control measures to prevent their spreading, intersected with other conservation issues. The crown-of-thorns became entangled in battles over exploiting the Great Barrier Reef for oil, limestone, and tourism.

4

UNDER CAPRICORN

From amongst the debate, opinions can be plucked at random and out of context. Motivations include idealism, financial gain, prejudice, personality clashes, and the scramble amongst scientists for scarce research funds.

But from any study of the Barrier Reef Controversy, one fact stands paramount over the scrambled battle.

Australians do not know enough about the problems posed by the Great Barrier Reef. Our research on it has been infinitesimal. As custodians of the greatest natural resource in the world, we have exposed nothing more than our own ignorance and rapacity. Our Governments, said an editorial in *The Australian*, are prepared to sacrifice the Great Barrier Reef.

To avoid involvement in this dispute is not easy. To reach a conclusion is impossible.

Vic McCristal, *Walkabout*, 1970

In May 1970, Chesher traveled to Hawaii, Guam, and Australia to discuss developments in *Acanthaster* control and research.[1] A mini-conference was arranged at the University of Hawaii, which had set up an *Acanthaster* research program. It was intended to embrace many aspects: population dynamics of adult and larval populations, analysis of larval development and survival in the plankton, analysis of metamorphosis, settlement of post-larval stages, predation of juveniles and adults, feeding preferences and rate, symbiotic relationships, the ecological impact of *Acanthaster* predation, as well as the feasibility of stimulating reef community recovery.[2] Joseph Branham and four or five colleagues at the University of Hawaii had been studying a population of about 20,000 starfish on Molokai. At the same time, the Hawaiian Fish and Game Division surveyed the Molokai population and carried out experiments using ammonium hydroxide to kill the starfish.[3]

In Guam, a control program established by the Guam Fish and Game Department was underway to monitor *Acanthaster* populations and kill large herds; a second program focused on coral regeneration and recolonization. Future plans included studies of changes in the fish populations and of the use of pathogenic organisms in *Acanthaster* control. The University of Guam would make its new marine laboratory available to *Acanthaster* researchers. It also planned construction of two trailer laboratories that could be set up anywhere on the island so researchers could live and work at the field sites.

The U.S. Trust Territory also maintained a control-and-monitoring program working in close cooperation with the University of Guam. Teams of divers were trained in Saipan and Truk, and diving lockers and boats were located at all district centers. Divers surveyed the reefs for large starfish populations, which they killed by injection with ammonium hydroxide, and simultaneously looked for valuable marine resources. The program also supported research by providing monitoring data and diving equipment. Each island was to be sectioned into permanent stations, and *Acanthaster* populations and other features of the environment recorded several times per year to provide a baseline for future studies of the dynamics of the coral-reef environment. Records of *Acanthaster* observed or "removed" would be kept. "Basically," Chesher commented, "the program is a holding action while researchers discover what, if anything, can be done in terms of permanent control methods, or if permanent control measures are necessary." [4]

In their discussions, Chesher noted that some scientists still spoke with caution about controls—that they were unnecessary or perhaps even harmful. But as Chesher commented, *Acanthaster* was not present at all in the coral reefs of the Caribbean and they did very well without it. [5] He compared the U.S. Trust Territory's *Acanthaster* control program to any other well-regulated fisheries, such as grouper or lobster. Also like those, the program was not so effective that it would endanger the species. [6] After all, if research later showed that the infestations were not natural and were undesirable, the reefs would still be alive. "Yet, if we do nothing," Chesher reasoned, "and let the coral die and later discover this was an error, nothing can be done to replace the living coral. Right now we have no guarantee that regeneration will occur and we certainly have no good ideas as to how long such regeneration will take. . . . While the problem is receiving scientific study, the islanders prefer to protect their reefs." [7]

In Australia, opinions over the crown-of-thorns problem were much more extreme and the political and economic context much more complex. Not only was the necessity of control programs questioned, even the reality of the infestations was doubted. In fact, Chesher was invited to testify along with several others, before a committee formed by federal and Queensland state governments, "to ascertain if there really is a starfish plague on the Great Barrier Reef." [8] The chair of the committee informed Chesher that "the committee will not set research priorities nor will it coordinate or recommend specific research projects; it is simply to find out if there really is an ecological problem on the Great Barrier Reef." [9]

Chesher arrived in Townsville off the central part of the Great Barrier Reef and met with Robert Endean. They made a brief trip to two reefs where Endean reported active, massive coral predation. The situation was as he described it. John Brewer Reef was "almost completely stripped of living coral. . . . The herds were massed much as they were on Guam." Lodestone Reef had large numbers of starfish feeding on living coral, and Chesher suspected it would soon share the fate of John Brewer Reef.[10] Endean later told him that the surveys "conducted independently by him, Robert Pearson, and the Department of Primary Industries showed about 180 reefs killed by *Acanthaster* which included most of the larger reefs between Cairns and Townsville, about one-third of the length of the Great Barrier Reef." [11]

Although Chesher had no reason to doubt the existence of the plague on the Great Barrier Reef, Endean's estimates of the magnitude of the plague, the risk to the Great Barrier Reef, and his call for action had been at the center of heated controversy for more than a year. The joint government crown-of-thorns Committee of Inquiry that invited Chesher to testify was formed after a fierce public controversy with environmental activists, scientists, and public opinion on one side, and the Queensland state government and commercial interests on the other. To understand the nature of that controversy, we must digress to outline the main events leading up to the formation of the Committee of Inquiry.

Cloak-and-Dagger Research

When Endean's original report commissioned by the Queensland government was released in June 1969, a year after its submission, press coverage of the *Acanthaster* problem dramatically increased. There was wide concern that no action had been taken. To many, this was another case of government reluctance to tackle the cause and effect of environmental abuse. A campaign for government action developed. In response to public outcries, a conference of government officers was held in Brisbane in August, to review the Endean report and Pearson's original survey. The meeting's conclusion was that "measures to control the starfish seem desirable, but no action should be taken until efficacy of any measures to be adopted had been clearly established." Further research into "the biology, ecology and population dynamics of the starfish" was needed.[12] However, the whole matter was thrown into confusion the next month when the Queensland premier, Johannes Bjelke-Petersen, announced at a crowded press conference that "the crown-of-thorns is not destroying the Reef." In direct conflict with Endean's report, the premier stated that there was "no real cause for concern. . . . Expert advice is that there is no vast plague . . . I have been shown evidence that coral on reefs has rejuvenated . . . We are not going to set up an elaborate enquiry into the problem." [13]

There was an immediate uproar in the press and on television, as those interested in the Great Barrier Reef began to divide into two separate camps. Some of the strongest denials of the significance or even existence of starfish infestations came from professional divers, who, as aquarium or tourist operators, had commercial interests in the Great Barrier Reef.[14] One of the "experts" who had briefed the premier was Ben Cropp. A well-known figure in Queensland, Cropp was a resourceful and experienced diver, spearfisherman, and cinematographer, whose films on wrecks, sharks, and similar adventurous topics were widely known. As he saw it, the starfish was probably helping rather than hindering the growth of corals. "We are coming to the conclusion that it is not a plague, and that the starfish wander in packs and play an important part in the natural growth of the reef." They do this, he said, by eating away the veneer of live coral on the staghorn species. As the coral dies, it collapses to form a strong platform on which the new coral grows. "In other words," he asserted, "the crown-of-thorns prunes the staghorn to give it a firmer foundation." [15]

Cropp's views were representative of a vocal body of opinion from the tourist industry that arranged itself behind the premier to challenge Endean's report and deny any ecological crisis on the Reef. The opposite view—that the only way to get anything done about the problem was to draw attention to it—was taken by very few of those involved in the tourism industry.[16] Although this might appear odd, the reasons for their denials soon became transparent to those engaged in crown-of-thorns research in Australia. Tourist operators seemed to fear that the disclosure that the most accessible reefs were destroyed would dissuade tourists from making the long and expensive journey to Queensland.

The experiences of Theo Brown are illustrative. In 1972, he wrote a little book about his involvement in crown-of-thorns research and the attitudes of tourist operators and politicians in Australia.[17] He had come to study *Acanthaster* by a tangled route, after teaching scuba div-ing, studying sharks, and diving for many years on various coral reefs in the Indo-Pacific. Brought up in Sydney, he moved to Perth where he joined the Western Australian Police Force and formed their first div-ing and underwater rescue squad in the 1950s. Subsequently, he moved to Darwin, joined the Northern Territory Police, and began their first scuba-diving club. In 1960, he was in Sydney on his way to New Zea-land to set up a police diving and rescue squad there. But those plans were aborted by tragedy. Before leaving Sydney he had gone swimming in Middle Harbour with two boys to teach them the use of mask, flippers, and snorkel. A shark attacked one of the boys, taking his right leg off above the knee. Brown managed to pull the boy out, but he died in the hospital.[18]

Consequently, Brown devoted himself to investigating means of pro-tection against sharks. Between 1961 and 1965, he conducted experi-ments on all known types of shark repellent in reefs around Magnetic Island, a tourist haven about 5 miles off Townsville. Initially, he sup-ported himself by running a charter boat service between the island and Townsville in Northern Queensland. Later, he continued his work in French Polynesia as an associate of the Medical Oceanographic Branch of the Institute of Medical Research of French Polynesia. On the reefs around Rangiroa, an atoll with a population of about 350 Polynesians, he found huge packs of sharks, forming a wall of about three hundred individuals. He experimented with sounds to attract and repel them.[19]

In June 1969, he worked as co-director of a joint United States–French expedition with Bruce Halstead, director of the World Life Re-search Institute in California. Halstead was a former naval surgeon and an authority on dangerous, poisonous, and venomous marine animals. Their aims were to evaluate the sonic approach to shark repellents, study behavioral problems and psychological reactions of divers in a shark-infested environment, and conduct research on poisonous and venomous marine animals in the area. Twenty U.S. scientists and assis-tants were in the team; the French made facilities available and gave other assistance.

By 1969, after twenty years of diving, Brown had seen only six crown-of-thorns. He first saw large populations when working with the team at Rangiroa. They were not in numbers that could be called an in-festation, but they were causing damage. However, he and Halstead heard of reports from the Great Barrier Reef that reefs around Dunk Is-land and those off Townsville were wiped out. Halstead was a member of the Great Barrier Reef Committee. Like Endean, he saw great possi-

bilities for extracting pharmaceutical products from coral reefs, and had
dived with Endean off Heron Island.[20] He had also studied
Acanthaster's venomous spines. In 1965, a year after Barnes and Endean
published their paper warning tourists about the starfish's spines,
Halstead carried out a study on the location of venom.[21] He sent Brown
to the Great Barrier Reef to investigate the infestation on behalf of the
World Life Research Institute.[22]

Brown arrived at the Great Barrier Reef in November 1969, setting
up base on Magnetic Island with two trainees and a ton of gear, including sonar and photographic equipment. He intended to conduct a survey and experiment with the transmission of sonic and ultrasonic sound
frequencies as a means of controlling starfish. There had been advance
publicity of his arrival and tourist operators put pressure on him the
very day he arrived. Brown recalled:

> They would say I was wasting my time as there was no starfish damage
> on any reef in the vicinity of Townsville. I was told this by people who
> visited the reef consistently with tourists. . . . Anyway, this initial pres-
> sure had the opposite effect on me. I have not served in two police forces
> for nothing. The moment somebody said to me, "Don't go out
> there—there's no starfish," I began to look for an ulterior motive. I
> would quote Dr. Endean's statement that Slashers Reef, for instance, was
> a write off. Someone would be sure to reply he had just recently visited
> that reef and found no starfish and no damage to coral, and that I would
> be wasting my time if I went out there. . . .
>
> This attitude annoyed me. . . . So I wondered what their angle was.
> Why were they trying to soft-pedal the problem to me—even though I
> had explained that mine would be an independent survey and that if I
> found the Crown of Thorns was no threat, I would say so publicly.[23]

The reason for their attitude became all too clear. They were frightened that publicity about the starfish plague would keep away overseas
visitors, and some made this complaint to him personally. Brown considered it a puerile argument: "If these people had sufficient intelli-
gence," he thought, "they would reap a bonanza over the next ten years
by advertising: 'Come and see the Great Barrier Reef, because soon
there'll be nothing left.'"[24] But not all tourist operators opposed his survey. Some offered him the use of high-speed cabin cruisers to make a
number of trips on the outer barrier reef in June and July 1970. He
would take a cloak-and-dagger approach, leaving Magnetic Island
about 2 A.M. and telling nobody where he was going. As he explained,
"I feared if the Queensland government learnt I was using charter run-
abouts for starfish research they might seek to make trouble for the

owners, as had happened already in one instance." [25] Brown observed mass destruction: "The plague spread south from Green Island to destroy the lovely coral gardens I knew. The Dunk Island reefs, and all those down to Townsville and beyond, have been devastated. John Brewer and Slashers reefs today are extinct skeletons of what once was a wonder of the world." [26]

FORGED IN A LARGER ENVIRONMENT

There was much more underlying the crown-of-thorns controversy in Australia than the inaction of government and the fears of those in the tourist industry about negative publicity. The exploitation of the reef for limestone mining and oil exploration had been a simmering conservation issue. Such explorations met with great resistance from environmental and political activists, especially among students at the University of Queensland in Brisbane. As Roger Bradbury, who devoted much of his scientific career to studying the crown-of-thorns, advises, to appreciate the context we need to briefly consider the politics of Queensland as well as challenges to bureaucratic authority. Bradbury explained: Queensland effectively had one of the first socialist labor governments in the world beginning in the 1880s. The state owned many enterprises; there were even state-owned butcher shops in Queensland. One can still see the facades in Brisbane, like "State Butcher Shop #63." [27] The view then was that the church, in the largely Irish-Catholic culture, would be primarily responsible for education. Everyone would have a basic education, to do basic manual work; high school would be only for the elite.

There were two leading parties in the 1960s and 1970s. The Labour Party was supported by the Australian Workers' Union: sheep shearers, miners, dock workers, manual workers. It was not trying to make a revolution but to keep things as they were, and with good pensions. Bjelke-Petersen headed the Country Party, the conservative party, and was premier from 1968 to 1987. The Country Party's unofficial slogan was "Capitalize your profits and socialize your losses." Because half the population lived outside urban areas, they had considerable sway. The Country Party also believed in strong government intervention. It was based on development, but country people needed government support all the time. "With the booms and busts of mineral commodity cycles, of sugar, wheat, wool, on which Queensland depended for its economy, everyone looked to government to take them out of the troughs." [28]

During the 1960s and 1970s, the University of Queensland was one of the hotbeds for political activism in Australia. It was also the university at which many of Australia's first generation of coral-reef ecologists were educated, including Bradbury. Its radicalization began around the time he arrived as a student there in 1964.[29] Just before Australia entered the Vietnam War, there were major issues about whether rugby teams from South Africa should be allowed to tour Australia. Those opposing the tour argued that the players were not selected on merit—they were chosen on apartheid, and to support them was to support apartheid. There were major demonstrations on campus, which before long embraced civil rights in Queensland itself. Bjelke-Petersen's administration passed laws to stop demonstrations and the apartheid issue was enlarged into an issue over the right to assemble. There were large marches into Brisbane from the university. Bjelke-Petersen declared a state of emergency to give police extraordinary powers. He bussed in police from country areas to reinforce the situation and they surrounded the university. The university went on strike.[30]

Thus, the University of Queensland was becoming very radical, one of the most radical universities in the country as Vietnam protests were just starting in Australia. As Bradbury recalls, "the vibes were in the Zoology and Botany Departments." Graduate students who went to the University of California sent pictures of demonstrations there. The upheaval began to be felt in biology as well. This was the time when professors in the zoology department turned away from teaching natural history to teaching ecology, mainly terrestrial. In Bradbury's view, all the issues—political and environmental activism and the turn to ecology—were related:

> You got interested in ecology, nascent environmentalism, so one becomes a justification for the other. Then you got this political awareness—suddenly a whole cadre of people are starting to think about the political contexts of their actions. We never did that before. When I started as an honours [student in ecology], it was complexity-diversity, write papers, understand these things in the purer sense. Suddenly, you're exposed to these environmental issues that were more complicated than you understood; they involved people and politics, social forces, and economic forces.[31]

In 1967, Bradbury was editor of a newsletter that later became the magazine of the newly formed Queensland Littoral Society. That year, activists supported by the Littoral Society, about seventy-five members, went to court to stop a lime-extracting company that had filed the first-ever application to mine coral on Ellison Reef. The company's claim was that the reef was dead. The Littoral Society sent a student to the reef to count the fish to show that the reef was not dead. It had lawyers who volunteered their time. It won its case and got some good publicity.

Then on January 1, 1969, the Queensland government opened the reef to oil prospecting and the conflict between conservationists and those wishing to exploit the reef began in earnest.[32] Already two exploratory drill-holes had been sunk and many more were planned. Eighty percent of the area inside the continental shelf protected by the Great Barrier was held under exploration permits by various oil companies, including one, Exoil-transoil, in which Premier Bjelke-Petersen himself was said to have a substantial interest.[33]

The Littoral Society began to raise money by selling "Save the Barrier Reef" bumper stickers. They sold thousands of them. The warnings of conservationists were soon strengthened. The risks of drilling were suddenly and dramatically publicized by a series of massive oil spills in the Santa Barbara Channel in California, in mid-January 1969. For weeks, newspapers and television newscasts showed blackened beaches covered with dead fish and volunteers struggling to save oil-drenched birds and dying seals. Public disquiet in Australia rose markedly, with questions in parliament, letters to the newspapers, and stern editorials. The cause of protecting the reef was joined by students and scientists, divers and surfers, wildlife societies, and the powerful trade unions that announced a pivotal boycott on drilling in the reef in 1970. The movement, backed by a largely sympathetic press, was led by prominent figures like the poet Judith Wright. Their weapons were opinion polls, scientific evidence, the courts, "Save the Barrier Reef" stickers and newspaper editorials.[34] Bradbury recalls:

> We started this pressure campaign. We suddenly learned how to do it. You do the scientific advice, you do the expert appearances, but you mount a publicity campaign in the press. The stickers put pressure on the politicians, that was the real thing. After that came the crown-of-thorns. Not only did the issue build, but you had this group of scientists becoming politicized. They're getting forged in a bigger university environment that was very radical, learning how to apply those activist tools, and learning to apply them to a conservation issue.[35]

The conflict came to a head with a symposium in May 1969 held by the Australian Conservation Foundation in Sydney. The symposium was planned to cover all aspects of conservation, but the potential risks of oil drilling captured most of the attention. W. G. H. Maxwell, an eminent geologist from the University of Sydney, argued that the great viability of the Great Barrier Reef "should not be obscured in the hysteria of ignorance," that "the possibility of excessive reef destruction by animals—human and non-human—would appear to be quite remote." [36] Maxwell's views were regarded with reserve by many conservationists because of his own financial links to oil interests. Critics pointed out, for example, that in his important *Atlas of the Great Barrier Reef,* published in 1969, he acknowledged that from 1965 he had received "extremely generous support from the American Petroleum Institute, and the Petroleum Research Fund of the American Chemical Society." [37] Maxwell also became a major proponent of the view that the crown-of-thorns infestation was a natural, cyclical phenomenon that would correct itself. In fact, he offered some of the first geological evidence in support of this claim. He also was a prominent member of the crown-of-thorns Committee of Inquiry that invited Chesher to Australia. However, one other crown-of-thorns committee was formed before the Committee of Inquiry was painfully born the following year.

ACADEMY OF SCIENCE COMMITTEE

As public disquiet about *Acanthaster* continued to mount, the Great Barrier Reef Committee suggested to the Australian Academy of Science in Canberra that it conduct an independent inquiry. In November 1969, a small ad hoc committee was formed. Its mission was simply "to consider the reported widespread destruction of the Great Barrier Reef and to report to Council." [38] Was there a plague or were Endean and the press exaggerating? If the former, what caused it, and what if anything, could or should be done about it?

The Academy's committee was composed of a member of the executive of the Commonwealth Scientific and Industrial Research Organization (CSIRO), the major national research body, a veterinary physiologist, a geologist, a geophysicist, a parasitologist, and an entomologist. It was chaired by a human geneticist, R. J. Walsh. The absence of any qualified marine biologists reflected the impoverished state of marine science. Indeed, there were few marine biologists conducting research in Australia and only about six had any direct involvement with coral reefs. [39] Although Australia has a coastline as long as the equator, washed by three oceans, it did not have one marine research center of world standard. The only scientific base on the reef itself was the research station on Heron Island that the Great Barrier Reef Committee had established in the 1950s. And the scientists who worked there complained about the lack of accommodation and equipment. For many years, Australian scientists had advocated the need for a national scientific center devoted to marine science. In 1970, there was news that the government would allocate $A3 million to erect a national institute of marine science slated for Townsville.[40] But, it would be another decade before the Australian Institute of Marine Science (AIMS) was completed and properly at work.

The Academy of Science's committee circulated word that it would meet in Sydney for two days in November 1969 to receive written and oral submissions. It obtained written submissions from Endean, Noel Haysom, Department of Harbours and Marines, Donald McMichael, New South Wales National Parks and Wildlife Services, Frank Talbot, director of the Australian Museum, Sydney, and Peter Woodhead who was then ending a three-year spell as research director of the Heron Island Research Station to take up a position at a new Marine Science Laboratory at Memorial University, Newfoundland. (He later succeeded Goreau as director of the Discovery Bay laboratory.)

The committee released its report to the press promptly a few weeks later. It confirmed that there was indeed a "plague of *A. planci* widespread throughout the Indian and Pacific oceans" and that there had been "serious destruction of live coral in scattered areas throughout these oceans."[41] It judged the surveys of Pearson and Endean to have been satisfactorily undertaken with the limited resources available.[42] Although less decisive in regard to the cause of the plague, the committee suggested it might well be natural. It pointed to preliminary data offered in evidence of periodic increase in the past. In a personal communication to the committee, W. G. H. Maxwell reported that his laboratory had reanalyzed sediment samples taken during a marine geological survey between 1963 and 1967 on the Great Barrier Reef. His researchers analyzed samples specifically for echinoderm content and found "an abundance of echinoderm remains at various age horizons." The committee recommended that these studies be continued to see whether *Acanthaster* remnants were present in the samples.[43]

Most committee members doubted that overcollection of the giant triton was a likely cause. They suspected that events associated with the larval phase of the starfish held the key to the problem. But they rejected Chesher's hypothesis that destruction of coral by bombing, blasting, and other human activities had caused the plague on the grounds that such activities had not been widespread in the region of the Great Barrier Reef.[44] They also rejected the notion that a buildup of exotic chemical substances such as the organochlorine pesticides had caused a disruption in the coral-reef ecosystem. They further noted that some devastation had occurred in remote regions uninhabited by humans.[45]

The committee argued that changes in the environment that could trigger off a population explosion were "often subtle and seldom had a single cause." Perhaps a rise in seawater temperature had contributed to an increase in survival of starfish larvae. All agreed that ecological knowledge was meager, and the Great Barrier Reef was too much of a mystery for the population explosion to be understood. Research was needed on the biology of *Acanthaster*, its predators, and the physical environment of the Great Barrier Reef corals.[46] The Academy's committee recommended that a crown-of-thorns advisory committee be set up to determine priorities for a coordinated research program.[47] In the meantime, it recommended local hand-harvesting of starfish to protect tourist sites.[48]

The Academy's committee clearly contradicted the views of the premier of Queensland about the nonexistence of the plague. But to environmental activists the response from government was far from satisfactory, especially when compared to that taken by the American government. Even the committee's recommendation that control efforts should be limited to tourist areas was ignored.[49] Critics repeatedly pointed to the Westinghouse report to the U.S. Department of the Interior that rejected the view that the plague was natural.[50] They also ridiculed the Australian Academy of Science's offhand dismissal of the view that pesticides might be involved. (Two years later, the Academy undertook a similar sort of study on the safety of DDT, concluding that it was "no serious problem." A week after their pronouncement, the United States banned DDT.)[51]

What became publicized as the "Starfish Wars" continued to build as the massed ranks of *Acanthaster* moved from Green Island to less accessible reefs. Reporters told of how swarms of 10,000 to 20,000 giant starfish had devastated much of Australia's Great Barrier Reef, as well as Guam and Fiji, while some 20,000 starfish were found about 1300 yards away from Hawaii's offshore coral reef. "It's an incredible story, never before known in the history of marine biology."[52] Sometimes all the usual suspects were grouped together as being responsible for the plagues. Religious voices joined in environmental protests. The magazine *Awake!* compared the Australian Academy's emphasis on the natural cycle argument to the U.S. Department of the Interior report, whose findings "indicate that man has upset the reef's delicate ecological balance by more means than one. They are: (1) excessive radiation from atomic testing; (2) too much collecting of the rare triton shell; (3) too much dredging; (4) DDT pollution and (5) fish dynamiting."[53]

Newspapers kept the views of activists before the public. Endean was widely quoted as stating that the situation was only getting worse and that the Great Barrier Reef might be completely wiped out in twenty years unless something was done.[54] In April 1970, Theo Brown held a press conference to gather public support for criticism of the Queensland government for ignoring the dangers of the starfish and to raise funds for his research on *Acanthaster*. He had just returned from Tahiti, where a new infestation was detected. About one thousand volunteers heeded a call from French Polynesia's Fisheries Department to "wage war" on the starfish. Brown told reporters how the death of coral reefs may bring a sizeable proportion of South Pacific natives to the brink of starvation. He called on the metropolitan governments to take immediate action to prevent this from happening. Pollution was the most likely reason for the upsurge of the starfish, as he saw it. Even the smallest intrusion on small plankton and cell life on the reef, such as diesel fuel spills from an outboard motor, could affect the balance of nature. If the starfish larvae had a certain immunity to pollution, then they might grow unchecked by affected organisms that usually prey on them.[55]

A QUESTION OF MOTIVES

Confusion among laypersons, scientists, and politicians was massive. From the point of view of "the public," assessing the social, economic, and technical variables in the controversy was as perplexing as it was for scientists assessing the cause of the plague. As one commentator put it, "Views ranged from anger at official inaction on the one hand to apathy and ridicule at the prophets of gloom on the other. Anyone who wished to make a comment was assured of a paragraph or two in the newspapers."[56] Reporters pointed to differences in disciplinary approaches among scientists to understand why some saw the starfish populations as a dangerous threat while others regarded them merely as a passing nuisance. As the magazine *Pacific Island Monthly* (PIM) described the debate, in scientific circles the two poles resulted from differences in perspectives of zoologists and geologists:

> In Scholastic circles the argument appears to lie between the zoologist and the geologist. The zoologist takes the problems of the starfish, examines its effects and declares it must be destroyed and its natural predator built up. The geologist examines the reef as a whole and is more inclined to treat the problem as one within the natural balance of nature.
>
> From the geologists' point of view the numbers of starfish could be due to something as simple as a rise in water temperature in certain parts of the ocean.[57]

There were financial stakes as well. To the geologist, a human-induced starfish plague meant a threat to valuable research funding for mineral and oil exploration on the Great Barrier Reef because all
human intervention would be called into question. It was thus important to establish the natural cycle, with human activities seen as infinitesimal in relation to this huge geological structure. However, many of
those calling for action rejected the terms of the debate—that if the
plague was natural, then there was no need to institute eradication programs. The "foibles of Mother Nature," who for unexplained reasons
sometimes allows one species to assume plague proportions and then to
die out, were a small comfort to the fishermen, naturalists, and tourist
operators faced with a threat to their reefs.[58]

By mid-1970, the various management issues concerning the Great
Barrier Reef, from coral mining and oil drilling to the crown-of-thorns,
were coming to a head. The main protagonists and the central stakes in
the controversies were well summarized by journalist Vic McCristal in
the magazine *Walkabout.* The combatants included powerful financial
and political battalions on the one hand and vocal sectors of public and
scientific opinion on the other. The rifts went deep and splintered in
various directions. Politically, there was friction between both Country
and Liberal parties *within* the state and federal governments, as well as
between them. The Labour opposition in both Parliaments opposed oil
drilling. There were sharp exchanges between Queensland Premier
Bjelke-Petersen and the Labour Prime Minister John Gorton.[59]

A clash of interests was reflected in the split among scientists: "Zoologists want the Reef left alone; geologists wish to exploit it." As
McCristal saw it, among the larger public, conservationists, skin divers,
fishermen, and boatmen opposed exploitation. Others, anchored to the
development of the tourist industry, assumed that development was
progress and that progress is automatic and part of the Australian way of
life. Opinions among "reef people" varied from suggestions that scientists were trying to leverage research funds from unwilling governments
to comments that the crown-of-thorns was being used to mask the real
assault by oil interests.[60]

In researching the story, McCristal discovered that one could easily
make any kind of case from among the welter of public comment and
scientific report.[61] The crown-of-thorns outbreak could be defined in
three ways: as the disaster it was said to be; as normal to marine life as a
flu epidemic in humans; as necessary for the cleaning and regeneration
of coral growths. One fact seemed obvious—Australians did not know
enough about the problem. Only recently had they taken much notice

of what happens on the Reef. Outboard boats, skin divers, and scientists were all new insofar as the Reef was concerned. The amount of research had been infinitesimal.[62] It was obvious that one could not discern the truth after a committee investigation. "The hideous marine gangrene we risk with oil drilling, or the silent advance of a parasitic starfish, demand decisive action. The thousands of people asking for this are not irresponsible. Australia would be sick indeed if her people were unconcerned."[63] The battle for the Barrier Reef, McCristal argued, was valuable for at least two reasons:

> It has exposed the commercial and exploitative approach of quite a number of our Parliamentarians, for one. More importantly, it has thrown light on a new kind of Australian, who shows enough concern for his environment to be willing to fight for it with the same weapons used against it. Our leaders must learn to avoid the trap of attempting to attain economic advance regardless of the cost. A wrong decision over the Reef can lead us instead into a new and hopeless poverty—a poverty of environment.[64]

Time was needed. "A moratorium, during which the Reef is more closely studied by our scientists. Ten years is not too long a period in which to do what modern man finds hardest of all—just to leave it alone."[65] Money was also needed. "Not the ludicrous $30,000 yearly projects of the past and the shoestring efforts of dedicated amateurs, but money adequate to the size of the problems." The planned $3 million marine research institute slated for Townsville would go far in this direction. The urgency for oil drilling, McCristal suggested, "just *may* stem from realization that once research makes the facts available, the oil industry will have a lesser chance." Leadership was needed. The fight to defend the Reef had been financed by idealists, or from sales of car stickers, and at great expense in time and money to some talented people. "It is past time to forget political and financial manoeuvring," he wrote. "The Reef needs a statesman, an environmental statesman."[66]

McCristal noted how overseas interest in the Great Barrier Reef issues had been keen and intensely opposed to exploitation. In 1970, two prominent visitors warned Australians of the dangers of pollution and ruthless exploitation of the environment. The Duke of Edinburgh cautioned against the overexploitation of fish stocks in the southern oceans, lest they suffer the same fate as in many northern seas as a result of thoughtless overfishing. "As it is, the blue whale will almost certainly be fished to extinction within the next few years," he stated, and "international control of fishing must be introduced before it is too late."[67] Pierre Trudeau, the Canadian prime minister, raised his voice to assert that although anti-pollution measures raised the economic cost of living, they would in time pay priceless dividends. The way in which Western governments measured the size of their gross national product, he argued, was false and misleading. "It assumes that our environment is self-renewing and inexhaustible. We know that this is not the case." He commented that oil tanker accidents at sea were not "spills" but "monumental disasters." The fear of such an incident occurring in the unique geographical and climatic circumstances of the Canadian Arctic led his government to propose legislation for the protection of the delicate ecological balance there. "If Governments do not prevent repetitions of this sort of activity we are all in peril," he said.[68]

In January 1970, after a meeting with Prime Minister Gorton, Premier Bjelke-Petersen yielded to demands for a full inquiry into the question of oil exploration on the Reef. A Royal Commission into oil drilling was set up. All companies with exploration leases in the Reef area would be asked to postpone drilling until the inquiry was completed and its findings considered. Twenty-six permit areas were affected.[69] On April 7, 1970, following pressure from the prime minister, it was announced that the Queensland premier had agreed to a joint government committee of inquiry into the crown-of-thorns starfish.[70] This was the committee that invited Chesher to testify. Its deliberations were to form the basis of government policy on the crown-of-thorns for the next decade.

5

CROWN-OF-THORNS
INQUISITION

The list of members read more like the Mad Hatter's Tea Party than an Official Committee on the starfish problem.

Peter James

And when the Premier says there is no plague, it's about the greatest plague in living history. I think it's going to continue to devastate the bulk of the hard corals of the Barrier Reef.

Robert Endean

The crown-of-thorns Committee of Inquiry was announced to the press in April 1970 with the news that there would be three members elected by the Queensland government and three from the federal government.[1] The federal nominees were:

- R. J. Walsh, Chairman, Professor of Human Genetics, University of New South Wales. He had chaired the *ad hoc* crown-of-thorns committee of the Academy of Science months before.

- W. G. H. Maxwell, Professor of Geology, University of Sydney. Specialist in the geology of fossil reefs, and employed as consultant to oil drilling interests in the Royal Commission on the question of oil drilling on the Reef.

- D. J. Tranter, Senior Research Scientist, Division of Fisheries and Oceanography, CSIRO. His research was mainly concerned with plankton, generally related to the Tasman Sea environment.

The Queensland nominees were:

- J. M. Thomson, Ichthyologist, University of Queensland.

- J. M. Harvey, Chemist, and director of the Department of Primary Industries, which included the fisheries branch.

- C. J. Harris, Accountant and administrator of the Department of Primary Industries.

The Committee's mission was comprehensive and its budget effectively unlimited. It was to determine "whether the crown-of-thorns starfish constitutes a threat to the Great Barrier Reef, and if so, the extent of such threat; and if necessary, to determine what control measures and/or further investigations should be undertaken, indicating an order of priority, and an estimate of costs." [2] It was to solicit testimonies of any experts it wanted to address the issues. Formed under intense public scrutiny, the committee proposed that "in the interests of maintaining an environment conducive to proper evaluation of scientific evidence, the press and the public would not be admitted during interviews." [3] But pressure from the media ensured that the public was admitted during interviews. About one year after its appointment, the committee presented its report; it was tabled in Parliament, then published and released in March 1971.

The committee heard evidence from twenty-nine witnesses, and accepted written submissions from a further ten individuals and two organizations. It held meetings at a number of centers, and visited reefs to observe conditions. It also examined reports in the press and scientific journals. "To date, without exception," Chesher wrote, after giving testimony in June 1970, "their witnesses have told them very similar stories. We know enough about the situation to realize it is an important reef phenomenon (and a potentially undesirable one) but there are too many unknowns to reach conclusions about the important questions. There is a real need for research into this problem." [4]

All eight marine biologists interviewed agreed that the problem was serious and that control programs were needed. Some argued that the plagues were anthropogenic, others maintained that they were natural and cyclical. The bulk of support for the latter view came from amateur divers and tourist operators. [5] But it was also supported by Peter Woodhead. [6] Over the previous three years, he and Jon Weber had conducted surveys in thirteen areas of the Pacific, over a period of "500 man-days" of exploration. They reached the conclusion that the abundance of *Acanthaster* had been seriously underestimated in the past, not only because of its nocturnal habits, but also because it was "difficult to recognize by an observer above water." [7] They further suggested that recent plagues were isolated, apparently unrelated, local population explosions, whose timing was merely coincidental. It was possible, even likely, that they had occurred in the past and may do so again in the future. "Whatever the cause," they argued, "local *Acanthaster planci* plagues must eventually exhaust their food supply and disappear, permitting recolonization of the reefs by coral in the same way that cyclone damage is repaired." [8] But Weber and Woodhead still recommended control measures in areas where the destruction of coral would seriously affect the local economy. [9] Others said much more.

When Theo Brown gave evidence, he reprimanded the Queensland government for its failure to institute control measures when Endean had recommended them two years earlier. [10] He accused it, and some tourist operators, of withholding from the public information about the extent of the damage caused by the infestation and aligning itself with those who would exploit the Reef's natural resources regardless of the consequences. He further accused Premier Bjelke-Petersen of "actively resisting the formation of the joint Commonwealth–Queensland committee of inquiry." [11]

Members of the Committee of Inquiry were less than sympathetic to such charges. In fact, they supported the premier's views that the population increases were indeed cyclical, and that the claims of a massive plague on the Great Barrier Reef were gross exaggerations. They ridiculed the many articles that had appeared in the popular press for "the repetitive presentation of eye-catching speculations or exaggerations of the limited facts available." [12] They further dismissed the main conclusions of Endean's report to the Queensland government on the grounds that his sampling techniques biased his data. Endean had estimated that twenty-three of the eighty-six reefs surveyed by Robert Pearson were damaged, and that on certain reefs 99% of the coral had been killed. However, the sites for counting were not selected at random, and entire reefs were not surveyed. Therefore, the committee argued that the conditions Endean described for entire reefs were actually only extrapolations, based on "counts made on 2 or 3% of the area, selected for coral abundance and by location at the northern and southern tips." [13] They further charged that allowance had not been made for the proportion of dead standing coral normally occurring on reefs.

The committee also used some differences in interpretation between Endean and Pearson to further discredit his report to the Queensland government. Following its submission in 1968, Pearson had sent a memorandum emphasizing that the triton control hypothesis was not substantiated and that further research was necessary before they were ever released. He also disagreed with the condition of infestation existing on one of the reefs.[14] The committee relentlessly attacked Endean's interpretations, blaming him for inciting the public campaign for government action and condemning his statements to the press as false and inflammatory. For example, in June 1970, Endean asserted in a television interview, and in the press, that there was a population of 3,000,000 starfish on Lodestone Reef. The next day, he corrected this to 300,000. A few weeks later members of the committee visited Lodestone and traversed almost the whole rim of the reef. They were accompanied by Pearson who, using the techniques employed in the earlier survey, calculated a population of between 10,000 and 30,000.[15]

The main thrust of the committee's report was clear: The reefs of the Great Barrier Reef were not being ravaged by a plague of coral-eating starfish. Reports of enormous numbers of starfish on reefs between Cookstown and Townsville had been greatly exaggerated. Heavy damage to corals had occurred only on a few reefs; in any case the corals would soon grow back. The reader was informed that the available evidence pointed toward periodic cycles of such starfish infestations, and that the destruction caused by the starfish was part of the natural forces of destruction that were a prerequisite for building reefs. Further, the reader was assured that the starfish plagues had not affected commercial enterprises such as tourism or fishing, and that action to control the starfish infestations throughout the Great Barrier Reef was not warranted. Nonetheless, the committee recommended that substantial government funds should be spent to further scientific investigations of the starfish infestations.[16]

CHEERS AND JEERS

Politicians welcomed the report as a timely corrective to the sensationalist exaggerations in the popular media. Premier Bjelke-Petersen was particularly enthusiastic about its assessment that, he asserted, "clearly vindicates the stand taken by the Queensland Government." He went on to challenge the scientific competence of Endean and those who "sought to spread gloom." [17] The Prime Minister of Australia, William McMahon (who had not long ago displaced John Gorton), also welcomed the findings in Parliament. He quoted the committee's condemnation of articles in the press, and added that "generalized action" comparable to that taken on Guam would be "ineffective" on the Great Barrier Reef because of its size and complexity. He also announced that the Commonwealth and Queensland governments would allocate funds for research as recommended by the committee.[18]

The applauding and cheering in the halls of government was met with jeers, along with charges of foul play, from environmentalists and several scientists who had given testimony. If Endean and others had exaggerated the risks based on little evidence and overestimated the extent of the infestations, the joint government committee had surely erred in the opposite direction based on the same evidence. To many critics, it was clear that this was not the impartial committee that was called for—it was precisely the opposite. It was a committee that had made up its mind before the fact, and marshalled evidence to support its case, while ignoring or dismissing out of hand what did not.

Frank Talbot from the Australian Museum in Sydney was one of the first to criticize the committee for passing off opinions as hard facts and being overly confident and too reassuring. He had testified about the reality of the plagues. He and his wife Suzette had recently published a paper concluding that although "opinions differ about the full extent of the damage and the possibility of ultimate recovery, there is no question but that a major disaster has occurred and is continuing." [19] Yet the committee's report made no mention of it. By dismissing any signs of risks, they only served to exacerbate the controversy. "The committee's task," Talbot remarked, "was surely to make a cool scientific appraisal of a situation in which there were violently conflicting opinions and much heat. Yet it has in a number of cases made strong statements on a fairly light platform of facts—the very error that one had hoped it would squash." [20] He could commend only that the Commonwealth and Queensland governments had agreed to set up a research fund to study the problem. [21]

Other commentators would not be as kind. Peter Dwyer, zoologist at the University of Queensland, charged that the committee's report was "a travesty of objectivity and on more than one account an irresponsible document." [22] As he saw it, committee members showed little understanding of the ecological concepts that confronted them, were biased in their presentation of evidence or opinion, and failed to maintain any semblance of internal logic in presenting their findings. He offered examples. In one breath the report stated that "knowledge of the reef ecology is inadequate to permit a complete assessment of present and future problems concerning the crown-of-thorns starfish," and in another that "the crown-of-thorns starfish does not constitute a threat to the Great Barrier Reef as a whole." It contradicted itself again, Dwyer argued, when it recommended that public money be spent on monitoring *Acanthaster* populations, reef ecology, and experiments in local control of the starfish, while insisting that there was no cause for alarm in any of these areas: "Any government that allotted money on the basis of such a report would . . . be irresponsibly misusing public funds. (I agree that money should be spent on research in this area but I am unable to find a rationale for this in the Report as it stands.)" [23]

The committee presented no evidence that contradicted statements that the reef was threatened. Thus, its conclusion that the population increase was "an episodic event which may have occurred previously" had no meaning whatsoever in Dwyer's opinion.[24] To illustrate bias in presentation of evidence, he pointed to their treatment of Woodhead's data on Green Island, about which they wrote that recovery of coral was "not as advanced" in certain areas as in others. Yet the basis of this re- mark was Woodhead's observation that "this was one of the worst areas seen on Green Island with rather little evidence of permanent recovery." Furthermore, Dwyer asserted, every important statement or conclusion in the report was ambiguous.

That the report might have a direct relationship to the Royal Com- mission on oil drilling was not lost on Dwyer. This was especially evi- dent in the committee's conclusion that "*A. planci* will not interfere with the integrity of the reef as a natural ecosystem." If one generalized this conclusion, he reasoned,

> we could be led to accept a view that any partial disruption of the reef will not interfere with its integrity and, hence, that one oil spill would be, in the long term, of minimum consequence. In light of my opening criticism the only way in which the Royal Commission can act, if it in- tends to consider the relevance of *A. planci* to the questions that concern it, is to discount the Report and start again.[25]

Considering the report in the context of conservation policy-making, Dwyer condemned it for being "blind to what conservation might mean and irresponsible in the views it puts forward." Conservation in the committee's eyes seemed aimed solely at protecting those spots that enrich some pockets and at boosting the GNP.

Endean wrote a scathing review of the committee's report, furious about its disturbing conclusions and the attacks on him.[26] First he dis- missed its conclusion that "serious damage" was "limited to some reefs between Cairns and Townsville." He did so using the same grounds as the committee did when they debunked his own conclusions. Both claims were extrapolations. The problem was that only 10% of the reefs of the Great Barrier Reef had ever been examined. How could the com- mittee justify their statement about the limits of "serious damage"? How did they define serious damage, and how many reefs were "some reefs"? Contrary to the committee's statements, Endean asserted that "*available data* indicate that the bulk of the coral cover of *most* of the major reefs between Cairns and Townsville was killed by the starfish during the 1960s and that the destruction continues." [27]

Endean also had allies for his view whom the committee simply ignored. It failed to mention Chesher's statement in June 1970 that "John Brewer Reef, about 6 miles in circumference, had lost 95% of its living coral to the starfish attack." It failed to notice the well-publicized statement of Bruce Halstead, director of the World Wildlife Fund—and a member of the scientific advisory committee for President Nixon and the United Nations—that the crown-of-thorns starfish infestations on the Great Barrier Reef were the worst that he had ever seen.[28] It also failed to recognize that when the bulk of coral on a reef was killed by *Acanthaster*, the whole biota of the reef was affected. Instead, the committee stated that "there is no evidence that the reef fish are seriously affected, at least in biomass." But this was merely a game of words. Of course there was no evidence, Endean asserted, because no one had attempted to determine the biomass of the fish involved.[29]

Thus, he argued that the committee had simply "chosen to ignore the weight of scientific evidence available which reveals all too clearly the known extent of the damage caused by the starfish and has attempted to minimize the magnitude of the damage already caused to the reefs of the Great Barrier Reef by the starfish, and has attempted to play down the possible consequences of this damage and of future damage to the living cover of the reef." [30] On the other hand, the committee did refer to the results of Maxwell's sediment tests of echinoderm debris to support their view that the population increases had occurred in the past and were natural and cyclical. Endean debunked Maxwell's results as irrelevant. They did not distinguish *Acanthaster* spines from other debris.[31]

Endean highlighted what he and others saw as the hidden agenda: how the natural cyclic view, that nature would look after itself, was firmly tied to commercial development: "Perhaps the Committee sees the starfish as posing a threat to Australia's ability to exploit the Reef for economic purposes." [32] He further pointed to the meager evidence the committee used to dismiss human-induced causality. To give some idea of the threat posed by pesticides, he offered the following information on insecticides and fungicides sold in one year by the North Queensland Tobacco Growers Co-operative Association (during the 1968–69 tobacco season):[33]

DDT (25% concentrate)	14,000 gallons
Endrin	7,000 gallons
Azenophos ethyl	2,300 gallons
Maneb	59,000 pounds
Matacil	1,000 pounds
Methyl parathion	470 gallons
Thuricide	194 gallons
Zineb	1,000 gallons
Linnate	3,000 pounds
Dithane	15,000 pounds
Cyclane	750 gallons

There was no evidence for any causality between pesticides and the starfish plague. But lining up potential suspects made it easy to imagine how they might eventually find their way to the sea and accumulate in the predators of starfish.[34]

Of all the statements in the report, the highlight for Endean was conclusion No. 12—"The committee is of the opinion that knowledge of reef ecology is inadequate to permit a complete assessment of present and future problems concerning the crown-of-thorns starfish and related matters." He found it incredible that a group of scientists holding this opinion and recognizing the "unique importance" of the Great Barrier Reef could have made statements that minimized the potential threat presented by the starfish plagues and did not recommend adoption of control measures while further research was carried out.[35] Arguing that the committee had failed to fulfil its "Terms of Reference," Endean called for the establishment of a Royal Commission to investigate all aspects of the starfish plagues.[36] In the meantime, he placed the responsibility for further loss of the coral cover squarely on the shoulders of the committee members and those members of Parliament who endorsed its report.

The joint government crown-of-thorns Committee of Inquiry was to have an influence far outreaching the dubious scientific basis of its findings. With its unequivocal assertion that "the crown-of-thorns starfish, *Acanthaster planci,* does not constitute a threat to the Great Barrier Reef as a whole," it set the Australian government position for a decade. Public interest in *Acanthaster* outbreaks waned. During the 1970s, most countries in the western Pacific region had taken action to prevent damage by the crown-of-thorns to coral reefs in their regions. Yet Australia had made virtually no attempt to control the starfish infestations on the Great Barrier Reef.

The Prickly Pear

Two books were published in Queensland in the 1970s dedicated to showing the official distortion about the crown-of-thorns. Both condemned the authorities and particularly the Queensland government for failing even to attempt some form of starfish control, for misleading the public, and for being aligned with those who wished to exploit the Reef—oil seekers, tourist operators, spearfishermen. Theo Brown's *The Crown of Thorns. The Death of the Great Barrier Reef?* was one. The other, *Requiem for the Reef: The Story of Official Distortion about the Crown-of-thorns Starfish* (1976) was written by Peter James. It centered around a critical exposé of the transcripts of evidence taken by the Committee of Inquiry and offered an insider's account of the dialogue and nature of expert testimonies. "Bias in intent is inferable," James wrote, "bias in presentation of the report is obvious." [37] He also emphasized how the Queensland government set out to destroy the credibility of those individuals, such as Endean, who stood against it. His aim was to set the record straight.

James pointed to the committee members' obvious lack of technical expertise and how, as a result of their incompetence, their inquiry lumbered along inefficiently through what was little more than banal logic, such as, "Well, anyway, I think the thought I had yesterday was: If the reefs were flourishing before the present infestation, and if there have been previous infestations, then there's been a complete recovery." [38] He further aimed to show how the committee attempted to steer witnesses to give answers that it deemed acceptable: "Do you regard this plague as not having done as much damage as sometimes claimed? . . . a passing phase . . . a cyclical phenomenon, perhaps?" [39]

Committee members often compared the starfish infestations to cyclical population explosions of rats, mice, oysters, and lemmings. However, respondents dismissed the suggestion. Michael Day, member of the Executive Board of the CSIRO, stated that when such population explosions were cyclic, they occurred relatively frequently. For example, rat and mice plagues rise in a year and die off. Yet, if the starfish plague was cyclic, its period of occurrence must have been a long one. A better analogy, he thought, would be the prickly pear menace in Queensland. [40] The committee offered the analogy again to zoologist William Stephenson: "The phenomenon of good and bad brood years is well known, for example, oysters, lemmings." Stephenson interrupted: "It is difficult to equate oysters with lemmings. The lemming explosion is regular at short terms and virtually predictable. The crown-of-thorns appears to have grown up on itself. If you want to draw a parallel, the prickly pear would be better." [41]

The damage to settlement and grazing lands in Queensland due to the introduction of the prickly pear cactus was well publicized in ecology texts as an environmental disaster that took years to control. [42] With the colonists' usual habit of carrying plants and animals into a new country, Captain Arthur Phillip had brought various species of cactus into Australia about 1787, intending to use them in culturing cochineal insects for dye. Some of the cacti escaped from his garden and by 1925 about twenty species could be found growing wild. Having no natural controls in this new territory, they spread prodigiously. By the end of the nineteenth century the "prickly pear" covered ten million acres. This increased to fifty eight million acres by 1920, despite attempts to clear it. At least half this land was so densely covered as to be useless. A commission was set up to find a cure using biological control.

In 1920, Australian entomologists were sent to North and South America to study insect enemies of the prickly pear in its native habitat. After trials of several species, three billion eggs of an Argentine moth *cactobastis* were imported. They were bred and released in 1926. Ten years later the last dense growth of the prickly pear was destroyed and the once uninhabitable areas reopened to settlement and grazing. Control of prickly pear was a long and expensive process with many people involved: research committees and research teams, field inspectors as well as the full cooperation of the rural community. The story of the prickly pear, James commented, stood in stark contrast with Australia's lack of efforts for research and control of *Acanthaster*. [43]

Like many other activists, James positioned the crown-of-thorns plagues in the context of all the problems "man, at least industrial man" had inflicted on his environment. [44] Human migration to new environments had often brought with it species extinction: twenty million bison roamed the North American prairies in 1870, five hundred were left twenty years later and these were saved only by the greatest pressures on government. By the mid-1970s, most species of whale were suffering a similar fate. Many exotic strains were introduced to new lands and ran amok in the same way that Australia got its prickly pear and rabbit.

Certainly the issue of instituting programs to control the crown-of-thorns infestations was not a matter for scientific evidence alone. The crisis hinged on more than marshalling "experts" or the limited available evidence to suit one's interests; it also hinged on other values. What was acceptable risk? This was not an issue that could be resolved by the quantitative standards of science. But it would be false to perceive the conclusions of the Committee of Inquiry as solely representing the opinions of non-experts, who, driven by ignorance and ulterior motives, deliberately ignored the warnings of experts. Matters of ignorance, or the intent of that committee aside, leading coral-reef "experts" in other countries were not nearly so certain as Endean and Chesher that this was a human-induced plague or that controls were warranted. Outside Australia, the controversy was drifting away from "science by press release" and into refereed technical journals as a pile of scientific publications began to grow on the crown-of-thorns problem.

6

A TREE FELL IN THE FOREST . . .

It is excusable to grow enthusiastic over the infinite numbers of organic beings with which the tropics, so prodigal of life, teems; yet I must confess I think those naturalists who have descended, in well-known words, the submarine grottoes decked with a thousand beauties, have indulged in rather exuberant language.

Charles Darwin, *The Voyage of the Beagle* (1860)

The early 1970s saw many new reports of large herds of crown-of-thorns starfish and devastation of coral-reefs in widespread localities, from the Indian Ocean to Fiji, Tahiti, New Guinea, the Philippines, the Ryukyu Islands in Japan, as well as the Great Barrier Reef and Micronesia. The reports came from amateur and professional divers, local fisherman, fishery departments, and museum and university workers. And they were scattered in various scientific journals, local government and conservation journals, as well as newspapers. Indeed, widespread concern about the crown-of-thorns paralleled a growing interest in submarine life as scuba began to be widely used by amateur divers and marine scientists.

Scuba had come in early forms in the 1940s, but was not popular until the 1960s and 1970s.[1] A number of best-selling books including Rachel Carson's *The Sea Around Us* (1951)[2] recounted the obscure life history of the oceans and contemporary oceanographic studies. Jacques Cousteau's *The Silent World* (1953) was complete with pictures of sunken ships, beautiful coral, and his companion Frédéric Dumas breaking a spear in a

50-pound grouper, telling how he took 220 pounds of fish from the Mediterranean in one morning. A new generation of coral-reef scientists was born in the 1960s and 1970s. In addition to scientific expeditions to the tropics, longer-term studies began to be carried out in newly established tropical marine laboratories.

Historical timing was right for research interest in the crown-of-thorns. The population of tropical marine scientists grew steadily from the 1960s. The first International Coral Reef Symposium was held in India in 1969, when less than a hundred people attended, and subsequently held every four years. The second symposium was on a cruise-liner, *The Marco Polo*, that sailed out of Brisbane, Australia, in 1973 with about three hundred fifty reef scientists on board. When the third symposium was held in Florida, seven hundred scientists attended. Coral-reef studies were published in several new journals founded in the 1960s, including *Biotropica, Oceans, Advances in Marine Biology,* and *Micronesica*. In 1980, the International Society for Reef Studies was formed, bringing together geologists and biologists. A new journal, *Coral Reefs*, was launched to cater to this increasing output.[3]

The new marine laboratory in Discovery Bay, Jamaica, opened its doors in 1970, while the new marine laboratory of the University of Guam was under construction. The same year there was word in Australia that a national institute for marine science was slated for Townsville. In 1966, the Smithsonian Tropical Research Institute (STRI) in Panamá expanded its scope by establishing a marine science program with laboratories on both the Atlantic and Pacific coasts. Already in 1971, before the senators' bill for funding was approved, STRI submitted a detailed preliminary proposal describing its interest in studying *Acanthaster* on coral reefs in the eastern Pacific where outbreaks had not occurred.[4] In England, a group of scientists at Oxford and Cambridge known as the "Cambridge Red Sea Expedition" changed its name to the "Cambridge Coral Starfish Research Expedition."[5] With Maurice Yonge as its Principal Scientific Advisor, and Cambridge biologist Rupert Ormond as its "Scientific Leader," the group's research interests in the crown-of-thorns were representative:

> What is needed is basic research into the reasons why this starfish, only seven years ago very rare, is now reaching plague densities in many parts of the Indo-Pacific. Research too is required into means of checking its further spread and into eliminating it in areas where it has already destroyed vast tracts of coral, in one stroke threatening the very livelihood of many islanders and removing a major and economically important tourist attraction.[6]

The plan to study the crown-of-thorns in the Red Sea, where no plagues had occurred, arose from two sources. First, it emerged from the idea that understanding the factors that kept *Acanthaster* in check in non-plague areas might help to indicate what had happened in plague areas. Second, it came from a suggestion of the "Cambridge Underwater Exploration Group" of diving in the Red Sea—the tropical sea most easily reached from England.[7]

Controls in a Sea of Rhetoric

During the early 1970s, the cause and effects of *Acanthaster* infestations were widely debated in the pages of *Science*, *Nature*, and newly founded marine biology and conservation journals. Those biologists who advocated that the starfish populations should be controlled while research continued, repeatedly warned of the potential risks: decline in fisheries and tourism, erosion of the geological structure of reefs, and, as Chesher had predicted, perhaps the extinction of hard corals in the South Pacific. On the other hand, all their factual claims, predictions, and demands were challenged—not just by the crown-of-thorns committees in Australia—but by leading marine biologists in the United States. Even the evidence that there actually was a population explosion of the crown-of-thorns anywhere in the Indo-Pacific was seriously challenged.

Debates in the scientific press began in 1970 with sharp exchanges between Chesher and biologists at the Scripps Institution of Oceanography, La Jolla. They underscore just how little was actually known about the ecology of coral reefs and how easy it was to offer alternative interpretations of existing evidence.

William Newman was one of the first to challenge Chesher's assertions about the potentially disastrous effects of the infestations and his call for control measures.[8] Newman first met Chesher when he, Goreau, and others discussed the results of the Westinghouse expedition in La Jolla in November 1969. Chesher had impressed him as a bright, young, and enthusiastic scientist. But at that time, Newman seriously doubted that the "plagues" were due to human interference.[9]

Newman was ten years Chesher's senior with considerable experience studying coral reefs and tropical marine ecology before moving to Scripps in 1962. After completing his Ph.D. in zoology at Berkeley from 1954 to 1957, he worked in the Caroline Islands at the Pacific Island Central School in Truk, sponsored by the Department of the Interior. He had participated in several expeditions investigating coral-reef evolution, ancient reefs, and the history of sea-level changes.[10] He had seen *Acanthaster* many times in Truk, but not the large populations and destruction reported in the 1960s. However, he had observed the natural formation of rather large clusters of other invertebrates known to be normally rare and dispersed on coral reefs. Perhaps the crown-of-thorns behaved in a somewhat similar way.

In the late 1950s, Newman studied the behavior and distribution of barnacles that burrow in coral limestone. Barnacles are plentiful on the shores of temperate waters. They are so abundant that those who have studied them remarked that the present period may go down in the fossil record as the age of barnacles.[11] However, this was certainly not true for the tropics. And there was a paradox. Although few in numbers of individuals, there were about ten times as many species of barnacles in the tropics. Darwin had a partiality for barnacles and had studied them in detail.[12] He was the first to note that, for some reason, coral reefs were unfavorable homes to large populations. Newman was interested in why. It was generally believed that physical conditions in the environment limited their abundance as they did corals. However, there was more to it.

Occasionally, Newman would find great numbers of barnacles burrowing in coral limestone.[13] But he discovered them only when certain kinds of fish, which grazed on microscopic algae that grew on the coral limestone, were absent. The fish possess a variety of incisor-like teeth with which they rasp away at the limestone reef. In their rasping they removed substantial amounts of limestone. Perhaps the reason for large numbers of the crown-of-thorns was similar: Alhough generally low in numbers, and scattered throughout the reef, they might simply crowd together sometimes, under certain natural conditions. Such ideas hovered in the back of Newman's mind throughout the early 1970s.

Newman and Chesher remained on the best of terms while in La Jolla, avoiding any confrontation of views. Just a few months before, they had engaged in a public debate in letters written to *Science* magazine. The issues Newman raised emerged at the center of discussion over the next decade. He began by arguing that there was no compelling evidence that such epidemics were unprecedented or that they constituted any significant threat to reefs and their inhabitants.[14] This was no more certain than the claim that *Acanthaster* was undergoing a "population explosion occurring almost simultaneously in widely separate areas." This apparent observation was, in reality, an interpretation, and Newman suggested that it might be due merely to the relatively new use of skin diving and scuba equipment in making underwater observations.

What was new then was not large herds of *Acanthaster* throughout the Indo-Pacific, but unprecedented numbers of divers observing them. After all, the fact that *Acanthaster* ate corals had become generally known only a few years earlier, and its predatory nature was then publicized in the popular media and semipopular scientific magazines. Attention drawn to the phenomenon, Newman argued, brought in new reports almost simultaneously from throughout the better part of the tropical western Pacific. Therefore, he reasoned, it was entirely possible that the "epidemics could have been occurring sporadically all along, on numerous widely scattered reefs across the Indo-Pacific, without being noticed." [15] Newman searched through literature from earlier in the century for statements to show that *Acanthaster* had been commonly observed without attracting attention. For example, in the 1946 edition of *Reef Shore Fauna of Hawaii*, the noted naturalist C. H. Edmondson, at the Bernice P. Bishop Museum in Oahu, reported that although *Acanthaster* was rare in Hawaiian waters, it was "abundant" about Christmas Island in 2 or 3 fathoms of water. [16]

Newman further shook the ground on which Chesher had erected his own theory—that destroying coral by dredging and blasting led to an increase in starfish larval survival. This suggestion, he argued, was pure speculation, and involved unknown aspects of larval mortality and behavior. In fact, Newman postulated that the inverse was possible: an abundance of certain kinds of coral might lead to more starfish. But, if reef damage was really the essential cause of the plagues, Newman insisted that other comparable natural causes of reef destruction, such as typhoons, should result in plagues. He noted that the area of infestation on the Great Barrier Reef was frequently hit by typhoons. Truk and Guam lie in a major typhoon track, and Palau had recently been struck by an unusual typhoon. [17]

That *Acanthaster* predation could spell disaster for fisheries was also baseless, in Newman's view. To the contrary, he suggested that large starfish populations might actually be good for fisheries because fish would graze on the algae that covered the dead coral. [18] The claim that *Acanthaster* predation on coral would result in erosion of reefs was equally shaky. Newman emphasized that coral was not the only organism involved in reef construction. Although biologists refer to tropical reefs as "coral reefs," many other lime-secreting organisms are also involved in reef-building. Various kinds of algae form filler material, and one, *Porolithon*, is a principal binding agent and a significant mass producer of limestone. This alga, Newman asserted, is primarily responsible for forming much of the seaward face of exposed reefs and protecting them from the destructive forces of waves. And there was no evidence that this living system was subject to damage by the crown-of-thorns.

On the grounds that the starfish was a normal member of reef communities, and on a lack of evidence to show that anything unnatural was occurring, Newman argued that the fierce campaign to eradicate *Acanthaster* populations was ill-founded and perhaps irresponsible:

> It assumed that the outbreaks are unnatural and in need of control, even though *Acanthaster* is part of the normal reef community and therefore must play its role in determining the quality of the reef complex. This role is unknown; should it prove to be important, indiscriminate exterminations of *Acanthaster* would then be considered highly irresponsible acts. Although it may be expedient to apply limited remedial procedures, provided there is some assurance they will do more good than harm, it would seem more valuable to put most of our available resources and energy into studying and understanding the nature of the epidemics before suggesting drastic control measures. Fortunately at least such studies are now in progress.[19]

Was Chesher being irresponsible in heralding so assuredly a disaster and advocating control measures? Or was Newman acting irresponsibly by minimizing the risks? Chesher was quick to respond, refuting each of Newman's arguments.[20] First, he emphasized that the aim of his various papers was to stimulate scientific interest in the problem. Moreover, he was not acting alone. Some forty international scientists had participated in the Westinghouse survey in Micronesia. "It was their unanimous decision," he remarked, "that the problem was significant and in need of extensive research and limited controls to protect valuable reefs." [21]

The possibility of indiscriminate extermination of the starfish was a red herring in Chesher's view. It was simply absurd to suggest that human efforts could reduce the numbers below normal population densities. It was also unlikely that anyone would kill starfish in places where they were not doing excessive damage to the coral. "Therefore," he remarked, "I doubt that we need concern ourselves at this point with over-controlling the problem." [22] Insisting that the threat to coral reefs, coral atolls, and their inhabitants was real, Chesher argued that it would be irresponsible not to institute control measures:

> The idea that destruction of one-quarter of the coral reefs on Guam is normal or does not constitute a significant threat to reefs and their inhabitants I find impossible to accept—as did the scientists who visited Guam last summer. Within the next 100 or 200 years the Guam reefs

might reach their former level of development again. There is, however, no compelling evidence that they will or will not. . . . If we take no control action against *A. planci* and let it kill a coral reef, we must be willing to accept its loss for several human generations.[23]

The debate over the cause and effects of the infestations floated in a sea of scientific rhetoric. To Newman's statement that there was no compelling evidence that such epidemics were new, Chesher responded that no evidence existed that they had ever occurred in the past. Again, he emphasized that the size, growth rates, and community structure of corals killed indicated that such infestations had not occurred within the last two hundred years on Guam or Saipan. Chesher dismissed the anecdotal remark about *Acanthaster* abundance in Edmondson's book of 1946 as ambiguous and irrelevant: "To an experienced naturalist, five specimens in one spot could mean 'abundant.'" However, a few months later, Thomas Dana, one of Newman's graduate students at Scripps, published further historical evidence in *Science* in support of starfish epidemics in the past.[24]

Dana pointed to several early reports. Surveys in the Philippines between 1924 and 1938 stated that *Acanthaster* was "common among corals and rocks" in Port Galera Bay.[25] In 1931, Danish echinoderm specialist and field biologist Theodore Mortensen reported that *Acanthaster* "was found rather commonly on the coral reef at the little end of Haarlem off Batavi . . . crawling over the top of the madreporarian corals on which it feeds, sucking off all the soft substances, leaving the white skeleton of the corals to show where it has been at work." [26] During their relatively brief but intensive shallow-water studies in Palau just before the Second World War, Japanese biologists reported the species to be "very common" on rocky and sandy substrata.[27]

Dana insisted that the historical rarity of the starfish had been greatly overstated, and the possibility that populations had "occurred sporadically but naturally in epidemic proportions on widely scattered reefs had been too summarily dismissed." [28] He asserted that coral reefs were "as adapted to such catastrophic events as are certain terrestrial communities to fire," and that "more harm than good could result from indiscriminate use of control measures." A few years later, Newman and Dana would offer their own theory for the nature of *Acanthaster* populations and go much further in their criticism of the need for measures to control starfish populations.

Whatever merits we might attribute to Chesher's arguments for controls, in the early 1970s several other leading coral-reef scientists, including some who had participated in the Westinghouse survey, believed that control measures had gone overboard. Starfish hunts were conducted in many countries to stem the population explosions and eliminate the plagues: Guam, the American Trust Territory, Hawaii, Western Samoa, the Cook Islands, Tahiti, Fiji, parts of the Ryukyus in Japan, and the Philippines. In some places bounties were offered ($3 per starfish at the phosphate-rich island of Nauru). In Okinawa, a private foundation and local governments contributed funds for a bounty system: 20 yen per starfish. The crown-of-thorns infestations had become an important issue because of an International Ocean Exposition planned to be held there in 1975. The bounty system was maintained from 1970 to 1975 and administered by the Okinawa Tourism Development Corporation.[29] In Western Samoa, where a bounty of the equivalent of 8 cents American was established, 13,873 starfish were collected along a 5-mile reef by March 1970. In Micronesia, approximately 200,000 starfish were destroyed by teams of "starfish eradicators." On Guam alone, 44,000 starfish had been killed by 1972.[30] In Tahiti, eighty nine volunteer divers removed 3200 starfish from the lagoon off the districts of Punauia and Paea in a period of five hours in October 1969. Damage to the reef was severe. In January 1970, six hundred volunteers removed an additional 3000 starfish from the same sector.[31] Globally speaking, the relative strengths of research and control efforts seemed severely lopsided.

HAWAII, 1970

While environmental activists in Australia complained of the lack of control programs, scientists in Hawaii complained of the opposite problem. Despite Chesher's assurances, there was indiscriminate killing and it was preventing scientific research. In August 1969, a large population of approximately 20,000 starfish was noticed on the south coast of Molokai in Hawaii. It was a large band varying from a few meters to tens of meters wide and about 2 km long. Instantly, there was a lot of concern. Biologists from the University of Hawaii and the Bernice P. Bishop Museum scrambled to do a crown-of-thorns survey conducted between October 1969 and May 1970.[32] Their main interest was population dynamics. They wanted to count the starfish to find out population density and other aspects of the starfish's biology.

In October, a mile-long transect line was laid at the bottom by the State Fish and Game Division. This main east–west line was crossed every 250 yards (228 m) by lines extending north and south 250 yards on either side. Five surveys of the starfish were taken at two-month inter-

vals. Divers swam along these bottom lines, recording numbers of starfish. Two were geologists; they wanted to do a survey of the bottom.
They took a small boat and a depth recorder and made a survey of an
area 5-miles-by-1-mile where most of the population was. At about the
same intervals, individual starfish were collected from the aggregation
and examined aboard ship by biologists from the University of Hawaii.
Each animal was measured, weighed wet, and examined for sex and go-
nad state. Teams of divers also made estimates of species composition
and the amount of dead coral along the transect lines. A few tagging experiments were also conducted.

Unlike the situation in Australia and Micronesia, the starfish at
Molokai did not appear to be killing vast amounts of coral. They
seemed to be concentrated on the faster-growing branching coral,
Montipora, which consisted of about 5% of the coral cover in that area.
Slower-growing *Porites* made up more than 80% of the coral cover. In
some cases, specimens of *Montipora* were found over-growing the
Porites, and scientists conducting the study believed that *Acanthaster*
slowed the growth of *Montipora* and helped keep the balance of species
diversity in that area. Unfortunately, before they completed their studies, the aggregation was featured in a documentary produced by a Honolulu TV station. In response to the publicity, in April 1970, the State
Fish and Game Division attempted to eradicate the aggregation. In a
three-day effort, approximately 10,000 starfish were injected, each with
10 ml of household ammonia.[33] At the University of Hawaii, Stephen
Smith recalled the incident:

> Chaplan Lam, who was working for one of the television stations, de-
> cided it was a big deal and he convinced legislature to go out and wipe
> the population out. Legislature appropriated a small amount of money
> and a bunch of volunteers came together—mostly military—and armed
> with syringes full of ammonia went off to kill the crown-of-thorns. Well,
> what they really did was to destroy the population as a place to do popu-
> lation biology because they killed enough of the organisms so the popu-
> lation dynamics was screwed up. But they really didn't make much
> difference in the population. . . . All they really did was ruin the re-
> search.[34]

Certainly, researchers at the University of Hawaii were aggravated by
the loss of the research project as well as research funding through the
culling of the starfish. Smith doubted that large *Acanthaster* populations
were new, anthropogenic, and would have long-term deleterious effects
on coral reefs if left alone. Attention to the crown-of-thorns rode a wave
of environmental issues: "A first awakening of environmental sensibili-

ties," Smith called it. This was at the time that the Environmental Protection Agency was just getting off the ground and scuba was becoming popular. He considered an issue about observer effects: "Once people become aware of something, they see it. You know, you see it once and you see it everywhere." [35]

JEKYLL AND HYDE

Was the whole issue of *Acanthaster* outbreaks a matter of media hype, the result of a growing population of recreational scuba divers and marine biologists? Was it only a "plague" of divers, an outbreak of reported sightings? By the early 1970s, many scientists came to that very conclusion. The experience on Hawaii in April 1970 was telling. It not only showed how an eradication program had interfered with research, but offered a striking illustration that, at least in some cases, large populations of the starfish may not do significant damage to reefs. It also demonstrated that what counted as a plague could not be determined simply on the basis of observations of large herds.

Indeed, the behavior of the population in Molokai threw massive doubt on the existence of damaging plagues occurring simultaneously throughout the Indo-Pacific. Additional evidence that this was indeed the case came from Peter Vine when he announced in *Nature* in 1970 that "the recent outbreak of reports about *Acanthaster* may perhaps be explained by great public attention and increases in the number of divers rather than of starfish." He added, "There is no real evidence to support the view that human interference is affecting the phenomenon to any extent." [36]

Vine worked out of the Department of Zoology, University College of Swansea, Wales. He had participated in the Cambridge Coral Starfish Research Expedition in the Red Sea in 1970. The Rotary International awarded him a Travelling Fellowship to visit many areas of the Pacific. He also spent a year in residence at James Cook University in Townsville, Queensland. Between December 1968 and April 1970, he conducted surveys of *Acanthaster* populations on eighty-three widely distributed coral reefs in the Pacific. However, of those he investigated, he considered the starfish common only on the Great Barrier Reef and Fiji. Vine rejected the claim that the first plague on Green Island occurred in the early 1960s, asserting that "an experienced professional diver from Townsville" informed him that there were large numbers on the southeast side of Lodestone Reef in 1954. Similarly, fishermen in the Solomon Islands recalled "large concentrations of them" forty years earlier, when night fishing on the reefs there was very hazardous because of the numbers of *Acanthaster.*[37]

As Vine saw it, there was no need to postulate a real increase in the numbers of the crown-of-thorns in the Indo-Pacific.[38] There were only local variations in starfish populations. These, along with "occasional plagues" were to be expected and were often recorded in invertebrates with such life cycles. All that was needed was a slight fluctuation in larval survival.[39] He argued that a periodic increase in seawater temperature would hasten larval development and in turn lead to increased larval survival by lessening the time the larvae are exposed to predation while in the plankton. There was also experimental evidence indicating that *Acanthaster* larval development was inhibited at temperatures below 24°C.[40]

Ridiculing suggestions that fishers would lose their livelihoods, reefs would erode, and atolls sink gradually beneath the waves, Vine insisted that *Acanthaster* predation was beneficial to coral reefs both in terms of species diversity and in strengthening their physical structure.[41] Goreau had suggested in 1962 that *Acanthaster* predation may be responsible for preventing the framework of coral reefs from growing. But Vine supported the inverse hypothesis: "Far from having a deleterious effect on the development of coral-reefs, dense predation by *A. planci* may actually promote their development by killing protuberant colonies and thus encouraging firmer regrowth and cementation basally. Thus by providing new surfaces for settlement of planulae and development and more rapid growth of young colonies, predation by *A. planci* may lead to an acceleration of reef growth."[42]

Few ecologists believed that *Acanthaster* helped coral reef growth the way Vine envisaged. But many thought that the starfish benefited coral reefs by enhancing and maintaining coral diversity because of its preferential feeding on the faster-growing coral. Its predation prevented those species from completely monopolizing available space and thus made room for others. In 1972, James Porter pointed to a growing consensus that the crown-of-thorns had a "preference for the most numerous common, faster-growing species of branching coral in the Indo-Pacific: *Montipora, Acropora, Pocillopeora,* and other genera."[43] At that time, Porter was a postgraduate student at Yale University. He did predoctoral research with Goreau in Jamaica in 1970 and subsequently worked as a predoctoral student with the well-known coral-reef ecologist Peter Glynn at the Smithsonian Tropical Research Institute in Panamá. His doctoral research, completed in 1973, was on the structure of coral reefs on opposite sides of the Isthmus of Panamá.

Acanthaster was relatively scarce in the eastern Pacific. However, Porter argued that its small numbers there in no way reflected the major role it played in structuring the ecological community. His studies indicated that the starfish preferentially fed on the most common coral on those reefs: *Pocillopora damicornis.* By preventing this coral species from monopolizing the reef and allowing the co-existence of rarer coral species, he argued that it effectively increased species diversity. This view fit well with ecologists' conceptions of the role of "keystone predators."

The concept of "keystone species" was developed in 1966 by Robert Paine at the University of Washington, Seattle. In experiments in Mukkaw Bay, Washington, he showed that removal of another starfish (*Pisaster ochraceus*) from an intertidal area led to a sharp decline in the species diversity of certain bottom-dwelling invertebrates and algae. *Pisaster* preyed on mussels and kept their populations in check, thus allowing the opportunity for other species to persist.[44] In 1969, Paine referred to such predators as "keystone species." [45]

Although Porter was convinced that *Acanthaster* would increase coral diversity, he was less certain of its overall benefits to coral reefs:

> Consequences of the presence of an *Acanthaster* population on a coral reef are complex. On the one hand, the biomass of living coral is reduced. In terms of resisting storm damage, for instance, this is probably detrimental to the reef. Furthermore, *Acanthaster* may exert enough predation pressure on a small reef or on a coral patch to prevent either from developing further. On the other hand, presence of the starfish increases immediately diversity, which in terms of resisting disease, for instance, is advantageous (monocultures are usually considered to be highly unstable). It is hard to say, then, whether the reef increases or deceases in long-term stability because of the presence of *Acanthaster.* However, the antagonistic properties of factors such as predation are clearly evident as we try to reach an understanding of the meaning of stability.[46]

Others were more confident, asserting that the *Acanthaster* was always of benefit to coral reefs, even in the long run. In other words, although its predation was harmful to a few species, it was good for the ecological community as a whole. The notion that *Acanthaster* was merely doing its job and was one of "Nature's most effective ecological balancers" led one biologist to refer to it as the "Jekyll-and-Hyde" starfish.[47]

Assessing the novelty and reality of "plagues," the risks, and the need for immediate action was indeed a complex task. One had to consider the social and political conditions—an environmental awakening, new technology (scuba), ecological theory (i.e., keystone species)—as well as

different coral-reef communities (i.e., different ecological settings). Joseph Branham was speaking for many marine biologists when he raised
all these issues in 1973. He had become involved with the
crown-of-thorns during the Westinghouse survey, whose objective, as
he saw it, was "to seek out the menace so that it could subsequently be
eradicated." [48] In an article in *Bioscience* he told the story of how he and
his colleagues from the University of Hawaii had tried to carry out investigations at Molokai. "More effort is being expended to exterminate
A. planci than to understand it," he lamented. "Such activity makes it
difficult to observe undisturbed populations for very long." [49]

Eradication programs and the vilification of *Acanthaster*, Branham
argued, were too emotionally charged. The thought of coral reefs con-
jures up images of swaying palm trees and creamy beaches and arouses a
sympathetic response in most of us. "Our vision is of a beautiful, intri-
cate, and probably delicate balance of nature, that grew up from the
depths eons ago, through the slow accretion of minerals by coral polyps." This vision contrasted sharply with that of the venomous, spiky,
carnivorous starfish predator whose very name "crown-of-thorns" (or
"mother-in-law's cushion" as it was known in New Caledonia) conjured
up emotion. Add to that the alteration of some "beautiful living reefs"
to "ugly dead ones," Branham concluded, "it is no wonder that
thoughtful men have been led to jump to conclusions." [50] The so-called
"population explosion," he cautioned, may reflect only the recent discovery of a situation that had existed for a long time, but was overlooked until appropriate incentives to observe it had developed. The
reports of "plagues" coinciding with increased numbers of people de-
scending into the depths to view the wonders of the *Silent World*, the
concept of *Silent Spring*, and the inflammatory language to describe the
situation, he argued, had combined to motivate a zealous search.

The very expressions "infestations, plagues, and population explosions" were too biased to be very useful scientifically, in Branham's view.
"Only prejudice distinguishes 'infestations' and 'plague' from 'normal
populations.'" The various definitions were based on starfish population density and/or the effect the starfish have on the reef. Yet those factors were so variable from place to place that the definitions were on the
whole unsatisfactory.[51] The dense aggregation at Molokai was not destroying the reef because they were feeding selectively on a species of
coral that "must have been growing at about the same rate as it was consumed."

Even when the starfish did have a serious effect on reefs, he emphasized, the environment was not "dead"; it was merely altered.[52] There

was no evidence that such "dead" reefs would be reduced to rubble and it was unlikely that the death of corals on some reefs would expose islands to increased erosion. Economic disaster also seemed unlikely because "most island people derive little benefit from the tourist trade nor do they depend heavily on coral-associated reef fish for protein." On the other hand, he quipped, "the starfish eradication business does offer a new source of income to some islanders." [53]

"It is naive," Branham remarked, "to think that *Acanthaster* is the only (or even the major) cause of coral death, or factor limiting reef growth." [54] He enlisted Darwin in support of the dynamic nature and shifting fate of coral-reef communities. During his voyage on the *Beagle* in the 1830s, Darwin made detailed observations of the structure and distribution of coral reefs. The *Beagle* had stopped at Keeling Atoll in the Indian Ocean about 600 miles off the coast of Sumatra. There, Darwin wrote in *The Voyage of the Beagle*, he "was much surprised to find a wide area, considerably more than a mile square, covered with a forest of delicately branching corals, which, though standing upright, were all dead and rotten." [55] Darwin thought this was due to events associated with changing ocean currents. Indeed, he argued further that "It would be an inexplicable fact if, during the changes to which earth, air and water are subjected, the reef-building corals were kept alive for perpetuity on any one spot or area." [56] In his book *On the Structure and Distribution of Coral Reefs*, he emphasized how some animals fed on coral "as showing us that there are living checks to the growth of coral reefs, and that the almost universal law of 'consume and be consumed,' holds even with the polypifers forming those massive bulwarks, which are able to withstand the force of the open ocean." [57]

Maligned and misunderstood, the crown-of-thorns had become a political issue, in Branham's view. It was understandable therefore, he wrote, that the U.S. Congress passed a $4.5 million bill "aimed at controlling the *Acanthaster* situation." Yet the money was not forthcoming. He learned from Senator Fong of Hawaii that the bill was not funded before reorganization in the executive agencies rendered it invalid. It was resubmitted and, as of May 1972, was pending before the House Committee on Merchant Marine and Fisheries.[58] Branham sighed in relief:

> Now the time has come for us to view *Acanthaster* and its impact on reefs more objectively, and fill in the gaps in our understanding with appropriate research. There is enough doubt as to whether or not the *Acanthaster* situation is unnatural to warrant suspension of the uncritical destruction of starfish. There is apparently also enough time to justify

research. The research should emphasize the basic biology of reefs as well as *Acanthaster*. We have learned from sad experiences, as in the case of DDT, that it is best to understand the whole situation *before* meddling with it, than to face the consequences later.[59]

A NONPROBLEM

In 1972, Newman and Dana teamed up with Edward Fager, a senior ecologist at Scripps, to propose a new theory. It denied the main "facts" that all previous theories were supposed to explain. There were no increases in *Acanthaster* populations at all locally or globally—all such assertions were due to "premature speculation." [60] Instead, they argued that the observed populations were only herding aggregations, migrating in search of food following typhoons. Nothing unusual was happening. They based their claim on a number of arguments and observations in the literature.

First they exposed a gaping hole in the claim that one was witnessing any "population explosions" in the Indo-Pacific. If this were the case, one should expect to see large populations of juvenile starfish preceding the appearance of adult aggregations. Yet they found it striking that large populations of juveniles had not been observed in any reports.[61] Of course, it was possible that small juvenile starfish were simply difficult to observe hiding among the coral. Nonetheless, there was no field evidence to indicate an increase in larval settlement or in survivorship at any stage of the starfish life cycle.[62] Therefore, they argued, the claim that large populations of adults indicated a population explosion was an unsupported speculation.

They organized a number of other observations to corroborate their view —that the observed large aggregations of the starfish represented redistributions of existing populations brought about under conditions of food limitation caused by typhoons. They noted how difficult it was to locate *Acanthaster* when not feeding; they hide in the crevices, caves, and undersides of overhangs. One way of making the starfish conspicuous would be to increase their active search for food. Wave action from typhoons would break the delicate branching and foliate coral on which *Acanthaster* fed and the starfish would survive such storms by hiding in crevices and caves.[63] Afterwards, one would expect them to be extremely active as they came out of their crevices to search for suitable colonies of living coral. There was evidence that individuals were able to locate others feeding nearby. Such an ability, at times of limited food supply, could result in an actively coherent aggregation on the march into adjacent, less disturbed areas searching for food.[64]

As a first test of their theory they needed to correlate the timing of the sightings of herds with meteorological records of the occurrence of typhoons. Data from the U.S. Naval Oceanographic Office for the Pacific for the period between 1947 and 1971 indicated that typhoon frequency in the tropical Pacific increased significantly during the 1960s. The fit was even better than one might hope. In 1965, a new record was set for both the number of warning days and the number of supertyphoons (wind speed in excess of 130 knots) recorded in the Pacific. Guam was struck by a severe typhoon in 1962 and was affected by the close passage of another in 1963 and one in 1964. Both 1967 and 1968 were years of unusually high frequency for typhoon strikes in the Marianas. "In fact, Guam, Rota, Saipan, and Tinian were all declared major disaster areas in 1968 (U.S. Naval Oceanographic Office, 1947–1970) and were all reported later as being "infested" with *Acanthaster*." [65]

The fit was also pretty good for Australia. Severe typhoons crossed the Queensland coast between Townsville and Bowen in 1958 and 1959. The 1959 storm was the third most intense on record.[66] Dana, Newman, and Fager asserted that the first reports of large numbers of *Acanthaster* "originated north of Townsville as early as 1959." [67] If their theory was correct, then large *Acanthaster* aggregations would have occurred sporadically in the past. However, they argued that, "with the unusual frequency of typhoons in the 1950s and 60s, the increasing number of divers investigating coral reefs, and a rising concern over man's impact on the environment, it is not surprising that such rare events were more frequently observed during the last decade." [68]

Vine's theory that an increase in sea temperature caused local "population explosions" by enhancing starfish larval metamorphosis seemed to hold little water. Data from the U.S. Coast and Geodetic Survey showed a corresponding decrease in water temperature.[69] In other words, as typhoon frequency rose in this region during the early 1960s, sea surface temperature declined; as typhoon frequency decreased toward the end of the decade, sea-surface temperature began to rise. If this data was correct, the waters off the Great Barrier Reef should have been cooler than average at the time of the first reports of large numbers of *Acanthaster*.[70]

Thus, Dana, Newman, and Fager rested their theory "that large aggregations of *A. planci* represent active behavioral phenomena; that is, they are redistributions of existing populations which at some point in their recent history have been brought under conditions of food limitation." [71] They quoted Darwin from *The Descent of Man* (1871):

Apologia

False facts are highly injurious to the progress of science, for they often endure long; but false views, if supported by some evidence, do little harm, for everyone takes salutary pleasure in proving their falseness: and when this is done, one path toward error is closed and the road to truth is often at the same time opened.[72]

Dana, Newman, and Fager's theory captured immediate attention. Many came to believe that the large populations were not new, or caused by human interference. As Joseph Branham put it in 1973, "It seems likely that the 'new' aggregations resulted instead from increased association of adults, mass migrations of existing aggregations, or were just newly discovered." [73] In the American popular press, the heralded fears of "starfish plagues" died out as abruptly as they began. The *New York Times* declared in January 1972, "Threat of Starfish to Coral Ends." [74]

Almost immediately, some leading scientists drew on the controversy as a lesson about the false problems of environmental alarmists and how they interfere with the proper aims of science. No one expressed this better than the director of Scripps Institution of Oceanography, physicist William Nierenberg. Speaking in Seattle in May 1972 at a symposium on the water problems of Puget Sound, he told his audience about two problems that drove Scripps scientists "to distraction." One was the "sociology" of environmentalists and the other was the "nonproblems" they tended to concentrate on. "Nonproblems," he said, "divert and dilute" the attention of serious workers and the public from the real problems, and he had never seen this concept emerge in as pure a form as in the ferment of environmental discussion. "Among the numerous possibilities," he chose "the story of *Acanthaster* as ideal because the tale is now complete," ended by the work of his colleagues:

> The story began about five or six years ago when it came to public attention via several routes that some parts of some coral reefs in the South Pacific were killed, and obviously by the crown-of-thorns, which was voraciously eating away at the edges of the devastated areas. The story was picked up in the world press and magnified in popular magazines. . . .
>
> The story expanded to predict the disappearance of entire coral reefs. This was to be followed by the total destruction of the atoll because of the loss of storm protection maintained by the reef. This is a very condensed description of an intense agitation that swept everywhere—even to the halls of Congress, where over five million dollars were authorized to the Smithsonian Institution to combat this scourge—fortunately never spent!

At some point in this development, the problem impinged unpleas-
antly on the consciousness of two of the few researchers genuinely competent in this area. . . . It took several years of part-time work but they
were able to show that in the past, right up to the present, blooms of
Acanthaster that upset the balance between coral and the starfish were
common occurrences. . . .

We could stop here as an example of a "nonproblem" but the story is
worth finishing. The question that remained is what caused these perturbations? Just recently the probable answer was found—the tropical
typhoons. A one-to-one correspondence exists between the *Acanthaster*
proliferation and the impact of the storm at the same spot. Apparently,
the reef is locally damaged by the storm, and the starfish congregate on
the undamaged remnants to start the process which eventually sub-
sides.[75]

Nierenburg's account contains a number of errors and misleading
statements. Most important, the controversy was not simply a matter of
those who were "genuinely competent" and those who were not. Indeed, scientists on both sides of the anthropogenic—natural divide
claimed that opposing views were the result of a lack of expert knowledge. Each side had accused the other of exaggerations or turning assumptions into facts. Moreover, Nierenberg had confused his own
colleagues' arguments. The hypothesis put forward by Dana, Newman,
and Fager was not one based on *Acanthaster* proliferation, but
Acanthaster aggregation—that is, it was a matter of feeding behavior,
not population increase. The *Acanthaster* problem was not considered a
"nonproblem" by any "experts," including the biologists at Scripps.
There was no clear evidence that the observed large populations were
common occurrences in the past, nor was it at all certain that typhoons
causing food limitations and migratory herds were the answer. The controversy was far from over—it was merely beginning.

7

KNOWLEDGE AND ACTION

Those who remain silent when their observations point to environmental decay are the undertakers of the environment; environmental post mortems become their stock and trade. They measure and we weep.

Robert Johannes, 1975

Nothing was certain about the crown-of-thorns in the early 1970s. Were infestations occurring more or less simultaneously throughout the Indo-Pacific? Were they natural, or induced by human activities? How extensive were they? How soon would recovery occur? Did *Acanthaster* predation normally enhance coral diversity? All the evidence could be given alternative interpretations and all the hard facts made fluid.

We must look elsewhere and consider other issues to understand the attitude of the disputants. Some have already been mentioned: an environmental lobby versus commercial and industrial interests in the Great Barrier Reef, zoologists versus geologists, fear mongering to attract funds for coral-reef research. We could also point to disciplinary and institutional interests in Nierenberg's attempt to close the controversy based on arguments of his colleagues at Scripps and to detract from environmental problems in general.

But there is still another important issue to consider further: the values of individual scientists.[1]

STYLE AND AESTHETICS

Although some coral-reef scientists cautioned on the side of "proof," others advised on the side of environmental decay. As Stephen Smith put it, "For some it was more important to get across your ideas than to do it with good solid science."[2] Smith saw the concern over the crown-of-thorns and its eradication campaigns as inspired by emotion, not science: "If an ecosystem can be charismatic, coral reefs can be charismatic and so a lot of people get emotionally attached and let that emotional attachment blind them somewhat to good regular science. . . . A lot of reef people have come into reef studies because reefs are such beautiful environments—so some people come into coral-reef research for the wrong reasons. They don't come in because they have serious scientific interests in understanding the environment. They got there because it is a little cuddly thing."[3]

Smith himself had come to coral-reef studies by a different route. He was interested in the cycling of carbon and the metabolic production of calcium carbonate. He studied limestone as a graduate student at Northwestern University in Illinois in 1967 and subsequently moved to the Institute of Oceanography in Hawaii, where he completed his Ph.D. in 1970. He later developed techniques for studying growth rates of coralline algae. After working at the National Museum of Natural History, he returned to Hawaii in 1973. Two years later, he served as acting director of the Marine Laboratory on Eniwetok Atoll, one of the two sites used for atomic bomb tests. Smith's comments about the aesthetics of reef studies and the tension between serious scientific interests and environmental activism require some examination. To do so we need to situate the crown-of-thorns in the larger context of the broader aims and interests of coral-reef researchers of the 1960s and 1970s.

It does not take long when speaking with coral-reef biologists to learn why so many had come into reef studies in the 1960s. They were lured by the tropical oceans, the love of diving, expeditions, and life in the South Seas. They were not motivated by a competitive laboratory chase for the Nobel Prize (there is no Nobel Prize for ecology). Instead of following the contemporary fashion for ultraspecialization in laboratory science, they were steeped in the age of the great scientists of the nineteenth century. They were not reading the fast-paced, "golden-hands" world of molecular biology as portrayed in James Watson's best seller, *The Double Helix* (1968),[4] but rather Darwin's *The Voyage of the Beagle* (1860).[5] Some began their careers as university professors and left the classroom for full-time research. Others were full-time laboratory scientists who left for field research.

The early career of John Ogden, Director of the Florida Institute of Oceanography, is illustrative. He began as a population biologist studying butterflies at Stanford with the well-known population ecologist Paul Ehrlich in the 1960s. Ehrlich had become increasingly concerned with human population growth, as described in his best-selling 1975 book *The Population Bomb*.[6] The problem, as he saw it, had been laid out in Thomas Malthus's *Essay on the Principle of Population*[7] of 1798. Malthus argued that populations, if unchecked, would grow geometrically while their food supply would increase only arithmetically. There would be too many mouths for the world's food supply, resulting in starvation and disease. Neo-Malthusian views and their political implications continue to be debated among environmentalists.[8] But these were not the issues that immediately interested Ogden as a student. Did some animals have a mechanism for regulating their own population sizes? If so, how do they do it? Posing these kinds of questions, he was engaged in studies led by V. C. Wynne-Edwards on animal dispersion in relation to social behavior. Ogden carried out experiments on the response of butterfly populations to changes in size and how to artificially select individuals for sensitivity to population density. He was also immersed in the "new community ecology" on species distribution, structure, and diversity led by Robert MacArthur.[9]

Before completing his graduate studies, Ogden complained to Ehrlich that his butterfly research had taken him deeper and deeper into the laboratory, but that he preferred the outdoors. His enthusiasm for fieldwork began with his own great voyage. Around 1964, Stanford obtained a 130-foot steel-hulled schooner from a lumber baron in Oregon. It was transformed into a research vessel at Hopkins Marine Station in Pacific Grove. A trip around the world was planned to study plankton layers in the Indian Ocean and learn what the community of macroplankton was about. For Ogden, "It was an opportunity you couldn't turn down."[10]

After completing his Ph.D. in 1969, he was awarded a post-doctoral fellowship at the Smithsonian Tropical Research Institute, where he remained for two years investigating fishes that graze on algae. In 1971, he moved to St. Croix in the U.S. Virgin Islands to help set up a new tropical marine laboratory for Fairleigh Dickinson University, a private university in New Jersey. Ogden remained in St. Croix for seventeen years, first as "Resident Marine Biologist," then as Director for his last six years. His main responsibility was to teach undergraduate students who came from New Jersey for three-week programs or for a semester. At the same time, he worked on various aspects of coral-reef ecology. The 1970s were dubbed the International Decade of Ocean Exploration by the NSF (National Science Foundation), which funded various programs on the open ocean and coastal processes and dynamics. One program was concerned with ecosystem studies of seagrasses around the world. Ogden established St. Croix as one of the five-year NSF-funded sites.

Shortly after arriving at St. Croix, Ogden reached back to Stanford and invited marine biologists Donald and Isabella Abbott to the laboratory. (Ehrlich also visited St. Croix; he had taken an interest in coral-reef fishes and learned to dive.) The Abbotts initially went to St. Croix to teach a short course with Ogden, which they turned into a research course. They worked together for four years studying the natural history of parrotfish, grunts, and the black spiny sea urchin, *Diadema antillarum*. In 1973, when Ogden began to study *Diadema*, they were the most abundant organisms around, known to hide in the coral and travel off the reefs at night to graze on sea grass. *Diadema* grazing was shown to be the major factor in the formation of a band, called a "halo," of bare sand between the base of patch reefs and outlying beds of turtle and manatee grass in Caribbean lagoons.[11] Moreover, as they grazed on microscopic algae that reside on coral limestone, they were also important for keeping those algae in check.

Beginning in January 1983, a devastating outbreak occurred—99% of *Diadema* in the Caribbean were wiped out by February 1984. It was the most extensive and severest mass mortality ever recorded for a marine animal. It started at the mouth of Panamá Canal, and the cause was thought to be an unidentified pathogen possibly imported from the Pacific through ballast-water dumping. The problems caused by transporting foreign species of fish, snails, crabs, plankton, and other creatures through ballast water are legion. The massmortality of *Diadema* in the Caribbean was followed the next year with the now notorious infestation of zebra mussels in the Great Lakes and other inland waters. The mussels were thought to have been transferred from the Caspian Sea in the ballast water of a transatlantic freighter.[12] The importance of clamping down on such stowaway species that wreak havoc on local ecological communities is obvious.

At the same time, studies of their effects provide a probe through which to study large-scale ecological change. For some, the mass mortality of *Diadema* offered the rare opportunity to observe the process by which species extinction might occur by invasion of a new species.[13] To

Ogden and many others, it also showed clearly that the Caribbean functioned as a large marine ecosystem. That reality posed serious limitations on the significance of ecological studies at any one particular research site studied in isolation. Ecological meaning, predictability, and generalizations also had to be sought in the context of the Caribbean marine system functioning as a whole.

Ogden became ever more concerned with the deterioration of the tropical marine environments and with what he called "long-term environmental research related to management." [14] In the early 1980s he was a member of the Coastal Zone Commission investigating sustainable economic development in St. Croix. Later, he played a major role in helping to establish a network of research sites throughout the Caribbean for monitoring coral reefs and providing a general context in which to distinguish and investigate the interaction between natural and anthropogenic changes. Although this turn to environmental concerns occurred later in his career, when at STRI, in Panamá, during the late 1960s, he and many others were troubled by *Acanthaster.*

Environmental activism and aesthetics were not necessarily at odds with good science. However, in the late 1960s many young coral-reef ecologists were initially reluctant to engage in public environmental issues. The early attitudes and experiences of Robert Johannes are exemplary. An advocate for *Acanthaster* control programs in the early 1970s, Johannes was in the forefront of his field—a pioneer in the development of ecosystem studies of coral reefs and the physiology of coral—algal symbiosis. He also became very involved in problems of pollution, fisheries management of coral reefs, the limitations of Western science and Western ideology, and the value of traditional ecological knowledge.

The aesthetic factor played no small role in Johannes' attraction to tropical marine science. Brought up in Vancouver, he completed his master's degree at the University of British Columbia on competition between rainbow trout and shiners in British Columbia lakes. His major professor suggested three different universities he might attend for doctoral research. Johannes sent away for the application forms: "University of Hawaii's application had a palm tree up in the left-hand corner. I didn't even fill in the two other application forms. That's how I got into coral reefs." [15] He completed his Ph.D. in 1963, on the flux of radioactive phosphorus through marine diatoms, arthropods, and bacteria.

He subsequently worked as a postdoctoral fellow at the Institute of Ecology of the University of Georgia. After two years, he started looking for jobs and was offered many. But the University of Georgia made him an offer he could not refuse. He remained there for thirteen years. That part of the country may not have been aesthetically pleasing, but the intellectual atmosphere more than compensated. Eugene Odum was there. "The nicest person you can imagine, incredibly creative in his thinking, spinning ideas off all the time," Johannes recalled. "It was an enormously stimulating time to be there because ecology became a word that was understood, starting in the very early 1960s, and ecology became the rage, and we had the best institute of ecology arguably in North America at the time. There was tremendous creative ferment, and people from the medical school, landscape architecture, as well as microbiology, botany, and zoology were all involved. . . . A lot of bright people and no prima donnas." [16]

During the 1960s, ecology was in a revolutionary period, moving from a descriptive science to one based on a dynamic functional approach, examining productivity and ecosystem processes.[17] This progression had occurred earlier in the century. In 1935, British ecologist Arthur Tansley coined the word ecosystem to join the separate studies of plant and animal communities into the study of the "biotic community." Later, the ecosystem concept was understood in terms of exchanges of energy and chemical substances.[18]

Assessing nature in terms of producers (organisms that first capture energy in photosynthesis—i.e., plants, algae, and some bacteria), consumers (all other organisms), and cycles of energy, ecologists began to measure energy flow and circulation of water, carbon, and other essential elements in ecosystems. The idea that one might be able to manipulate these variables to maximize biological production lay in the background.[19] The famous Odum brothers, Eugene at Georgia and Howard at the University of Florida, were instrumental in developing ecosystem research. Eugene Odum's book *Fundamentals of Ecology*, first published in 1959, became the manual of ecosystem thinking.[20]

PROJECT SYMBIOS

In Georgia, Johannes lived on the coast and worked at the university's Marine Institute, studying the role of bacteria and protozoa in nutrient cycling.[21] But it was not long before he felt the charismatic tug of coral reefs. "Coral reefs are warm, beautiful, a great place to work, but I was also intellectually utterly fascinated by the ecology of coral reefs and the ecology of corals." In 1970, he and Lawrence Pomeroy organized a key expedition that led to a solid framework for understanding the "metabolism" of coral-reef communities: a major ecosystem study on coral-reef productivity at Eniwetok Atoll. A classic study had been done there in 1954 by the Odum brothers,[22] but their techniques for measuring productivity were crude. Johannes and his colleagues' aims were to repeat and extend those studies with more sophisticated methods. "Project Symbios" was carried out at Eniwetok in 1972 led by Johannes with about twenty others, including Stephen Smith.

One characteristic the Odums had pointed out was that coral reefs were among the most productive ecosystems. But this is an enigma because coral reefs are living in an oceanic desert; nutrients are relatively scarce. So how do they do it? Part of the answer was located in the symbiosis of coral and algae. The algae living in the transparent flesh of the coral polyp use the sun's energy to transform carbon dioxide and water into organic chemicals, give off oxygen, and promote the coral's growth. The algae, in turn, utilize the polyp's carbon dioxide, nitrates, and phosphates.[23] Although the algae–coral argument was not the full answer, it is a microcosm of that answer. At whatever scale ecologists look, coral reefs are very efficient at recycling material. As reef ecologists put it, "Coral reefs don't leak very much."

Johannes and Pomeroy discussed the idea of carrying out the scientific expedition to Eniwetok Atoll in 1970. At that time Stephen Smith asked if they were going to study the most important metabolic process on reefs: limestones. They had thought only of organic material—carbon, nitrogen, and phosphorus—because that is what biologists tend to think of. Smith knew little about phosphorus and nitrogen cycling, but he did know about limestone. He joined the Eniwetok expedition, learned some ecology, and examined both inorganic carbon cycling and organic carbon cycling.

The Eniwetok expedition was largely funded by the AEC (Atomic Energy Commission). It had a marine laboratory there, initially established for studies of the impact of radioactivity after atomic tests during the 1950s. Like the Office for Naval Research, the AEC funded a great deal of research peripherally related to any naval military need. Several scientists had NSF funds and the expedition had the support of the Scripps Institution of Oceanography's floating laboratory, *The Alpha Helix*. It was tied up at the dock and researchers used small boats to do their work. But the site was still in military use; unarmed missiles were fired from Vandenberg Air Force Base in California with the lagoon as a target. Scientists remained there from six weeks to two months.

The Eniwetok expedition was highly successful—a major turning point in coral-reef research. It clearly demonstrated how one could study a reef as a system, not just observing species, but examining energy flows. About thirty papers emanated from the project. Smith listed its achievements: "The first serious calcium carbonate budget for coral reefs was produced; coral growth bands were discovered; phosphorous cycling on reefs was understood; the whole notion of nitrogen fixation in a marine environment is largely traceable to that expedition—coral reefs fix enormous amounts of nitrogen." [24]

Such fundamental studies of ecosystems were not far removed from applied concerns.[25] Supported by public funds, Johannes argued, ecologists were charged with two important duties: the preservation of ecosystems and discovering how to extract optimum sustainable yields from food-producing ecosystems.[26] Both aspects applied to studies of coral-reef ecosystems. One was for the early detection of "environmental stress" on coral reefs. Just as medical doctors measure various components of an individual's excreta to monitor its health, so might ecologists some day monitor the health of whole ecosystems. Instead of having to wait for the usual signs brought about by structural damage caused by pollutants or other environmental stresses, coral-reef scientists would detect their "metabolic changes" and perhaps be able to diagnose ecosystem stresses and remedy them before any irreversible structural damage began.[27]

Ecosystem studies were important for the conservation and management of fisheries.[28] Although coral reefs are among the most biologically productive communities on earth, their fish populations seemed surprisingly vulnerable to overharvesting. Johannes tried to formulate a biological hypothesis to explain this. Only later did he recognize the importance of placing humans themselves in his hypothesis. In the mid-1970s, when he lived among fishermen in the Palau district of Micronesia, he became aware "of various political, cultural, and economic pressures impinging on fishing in such a way as to make any purely biological explanation seem quite simplistic." [29]

TRADITIONAL ECOLOGICAL KNOWLEDGE

Local fishermen taught Johannes the seasons, lunar periods, and locations of spawning aggregations of some fifty-five species of food fish. They knew more than twice as many species of marine animals exhibiting lunar spawning periodicity in their waters as biologists had described for the entire world.[30] Such "traditional ecological knowledge" (TEK), gained during centuries of practical experience, was invaluable for managing tropical fisheries, Johannes argued. But to embrace such knowledge, not to dismiss it as anecdotal, meant that one had to challenge any concept of science that defined it exclusively in terms of controlled experimentation and rigorous statistical testing. "The important criterion" he asserted, "is whether it provides us with understanding. To expand a biblical quotation: by their fruits shall ye know them—not by their roots." [31]

Western science as well as Western values overlooked the ways in which Pacific islanders managed their fisheries. The delicate reef fisheries had been maintained for centuries by various taboos and traditions, based on legends and religion, but most important by what Johannes called "reef and lagoon tenure." The right to fish in particular areas was controlled by a clan, chief, or family who regulated the exploitation of their own marine resources.[32] However, the importation of the Western tradition of "freedom of the seas" and capitalist economies had weakened or destroyed such practices in many Pacific island areas.

At the beginning of the eighteenth century most countries, islands, and territories of coastal reef areas had been colonized by Western countries. By implanting Christianity, the administration of the colonies eradicated local religious traditions and habits, which had preserved reef resources. Western peoples' attitudes about the sea differed fundamentally: such fish resources were not owned by anyone. They were free for all to catch as much as they could, and did so in the manner that Garrett Hardin called "The Tragedy of the Commons." [33] As fishermen began to avail themselves of certain technologies—outboard motors, diving equipment, reef-walking shoes—resources dwindled. The supplanting of traditional fishery by commercial interests soon led, in many regions, to the degradation of reefs, overfishing of valuable species, and a drastic decrease in fish stocks.[34]

TO ACT OR NOT TO ACT

Today, Johannes works in Tasmania as a consultant researcher pursuing his chief interests—to integrate TEK and customary fishing controls with scientific knowledge and improve tropical marine resource management and conservation.[35] Yet initially he was reluctant to get involved in environmental problems. As he recalls,

> When I was approached at the University of Georgia by a young woman to get involved in public environmental issues, my attitude was that I was interested in basic science, not applied science. Environmental problems were things for someone to do, but not me. I don't know what exactly changed me, but somewhere along the line, for some reason I don't remember, I changed my mind, and got involved in public issues periodically. I never became what I would call a systematic crusader, but periodically I got involved in public issues I felt I knew something about,

and more and more became interested in applied research, and after the Eniwetok expedition, came my year in Palau, when I became aware very much of the importance of people in these environmental equations, which as a typical academic I never gave any consideration to before.[36]

Before his studies of TEK, even before "Project Symbios," Johannes had become involved in public environmental issues. It began in 1970 when he blew the whistle on the pollution of Kaneohe Bay in Oahu. After remaining in Georgia for five years, he began to visit Hawaii regularly during the summers to investigate coral physiology and various aspects of coral biology. When he started going back he saw that at one of his study sites in Kaneohe Bay a "green bubble alga" called *Dictyosphaeria cavernosa* was growing all over the reef, where it once appeared only occasionally. It not only engulfed corals, but actually dissolved their bases. One could lift the living coral off the reef because it was just sitting on a cushion of this algae; it looked terrible and the water was murky. After speaking with his old professors and graduate students who had been there all along, he realized that the change had been gradual:

> It is like the way you learn that the moon moves across the sky—by looking at it periodically. If you fix your eyes on it, you can't see it move. Well, they couldn't see the bay change because they were there all the time. Well, I hadn't seen it for five years and I could see it was in bad shape. So I started a newspaper campaign, which initially my colleagues were very skeptical about, except Jim Maragos. He was a graduate student then, but he jumped on the campaign with me, and gradually we won other people over.

In articles entitled "How to Kill a Coral Reef" in the newly founded journal *Marine Pollution Bulletin*, he described the degradation of Kaneohe Bay due to nutrient pollution and excessive sedimentation smothering the coral reefs because of bad land management practices. More than 3.5 million gallons of sewage, receiving only primary or secondary treatment, poured into that bay daily.[37] Johannes went further in his public campaign. In Hawaii, in the summer of 1970, he met up with Lee Tepley, a physicist from California who for fun had built himself an underwater camera with time-lapse facilities and wanted to do something with it. Johannes recommended that it might be used to settle a major controversy over coral feeding and how much the coral actually relied on the symbiotic algae. They took time-lapse pictures day and night and showed that corals were eating very little zooplankton.[38]

Because their camera was doing all the work and they had a lot of spare time on their hands, Johannes suggested that they make an underwater film documentary on pollution. He directed and scripted it; they called it *Cloud over the Coral Reef.* It was crude, with a bad soundtrack from 1935 travelogue music, and edited to alternate pictures of beautiful reefs with devastated ones. They put the crown-of-thorns in it as well. The footage served its purpose. Within a year, their half-hour film had been seen by 25,000 Hawaiian school students (and shown on Australian and Samoan television as well). It also put the campaign to clean up Kaneohe Bay on track and helped Hawaiian researchers get the funds to show that the bay really was in trouble. Ultimately, $26 million was to be spent to get sewage out of that bay, and today Hawaii has some of the best legislation on pollution of seawater in the United States.

Two key biopolitical issues underlie strong environmental advocacy, in Johannes's view. First, most coral reefs are situated in developing countries with very limited conservation and research support.[39] Second, there were the ethos or values of institutionalized science, verification of facts, suspending judgment "until all the facts were in," and all that this entails. The problem was that environmental crises develop faster than they can be completely assessed. It was more important to make interim decisions *in time*, he argued, than to make more scientifically satisfying decisions later. He warned pollution biologists against "injecting too much traditional laboratory caution into matters of immediate practical concern." They hesitate because their information is not final: "The fact that biologists have sometimes erred in their warnings about environmental degradation is no justification for abdicating the responsibility to speak up—any more than the fact that scientists sometimes publish mistakes means we should abandon publishing." [40]

The same argument applied to crown-of-thorns infestations. As Johannes saw the situation, the starfish were certainly harmful to reef communities on a short-term basis, though it was not certain if they were ultimately beneficial over a longer period. But, it could take decades to answer that question. Insisting that an "interim decision should be based on what we know rather than on what we do not know," he reasoned that until new information dictated otherwise, serious *Acanthaster* infestations should be controlled irrespective of their origin.[41]

Like Chesher, Johannes denied any inherent logic in the argument that if the outbreaks were natural, no controls should be instituted and nature should be allowed to take its course. "Anyone really committed to this philosophy," he quipped , "would presumably allow his house to burn down if lightning struck it, rather than interfere with a natural phenomenon." [42] "Cancer is an analogous phenomenon—the over-growth of one part of the system at the expense of the rest. Few would seriously argue that this natural phenomenon should be allowed to run its course." [43] Johannes's views, of course, were in conflict with those who believed that *Acanthaster* predation was beneficial to coral reefs even in the short run. But the latter claim would be refuted by closer studies in the early 1970s.

Chesher, like Johannes and other marine scientists, revised his approach to science following exposure to severe environment degradation. "Scientific objectivity," he later commented, "seems pretty frivolous when one returns to a coral reef that was once a thriving, diverse, and beautiful life system to find it battered into rubble by enthusiastic island fishers. There comes a time when one must combine scientific investigation with resource management." [44] After heading the Westinghouse survey, Chesher worked for WORL for two years; during that time he investigated the impact of a major desalination plant on the nearshore marine ecosystem of Key West, Florida. The Environmental Protection Agency (EPA) was just getting going then and tried to recruit him away from WORL. However, the EPA would have him working in an office and laboratory complex in North Carolina, and Chesher had not become a marine biologist to sit in an office. A career in academia was unappealing for the same reason.

Acanthaster had changed his priorities. As he recalled, "When I was in graduate school, the issue of human impact on the marine environment did not exist. Virtually no marine scientist considered humans could seriously impact ocean ecosystems." [45] Yet, in his view, *Acanthaster* showed people to be the most critical control species of island ecosystems. He had a practical goal of learning how to go about changing what people were doing to marine environments. This was the main problem introduced by the environmental movement as a whole. Information is crucial, but another central and very difficult problem remains: virtually every instance of environmental improvement asks people to do something that is against their immediate self-interest. Westinghouse, of course, was not in the business of changing people's behavior toward protecting the environment. It indulged Chesher's interest in *Acanthaster* and later coral-reef studies, but its corporate interest was how to make sure its various engineering projects did not run afoul of new environmental legislation. Westinghouse wanted an environmental engineer. As an agent of social change, the EPA was hardly effective, in Chesher's view.

In 1972, he bought a large research vessel, an 18-meter aluminum motor sailer, and continued environmental research in the Florida Keys and the Bahamas as vice president of the Marine Research Foundation. Between 1972 and 1975, he conducted sixteen Earthwatch expeditions using volunteer divers to help document the impact of a variety of human activities on marine ecosystems—dredging and filling, power plants, leechates from public landfills, and anchor damage. By 1975, he had had enough of swimming over impoverished habitats, showing up in courtrooms as an expert witness, and working on engineering problems. In 1976, Walter Starck, the founder of the Marine Research Foundation, invited him to join him in the Solomon Islands where Starck planned to establish a small research station. Chesher arrived in the Solomons in 1976. He had sold his large research vessel and bought *The Moira*, a 44-foot cutter, and engaged in various environmental, scientific, and political programs. He worked for several years as a consultant to the South Pacific Regional Environmental Program. He also worked on several projects with the United Nations Economic and Social Commission for Asia and the Pacific when they fit his own programs. In the early 1980s, he spearheaded a successful effort to end dolphin shows in Australia. Later, he worked with Earthwatch volunteers to help convince villagers in the Kingdom of Tonga to set up community-run giant-clam sanctuaries. In 1995, he organized an environmental educational program called Sea Keepers in New Zealand with more than 1250 participating schools. Currently, he is director of Tellus Consultants Ltd., a group of scientists advocating community and student participation in field research.

The Plagues Are Real

In 1973, while still in Key West, Chesher and Endean coauthored a response to assertions that their original claims about crown-of-thorns plagues were greatly exaggerated.[46] The events in Molokai, the reports of Porter, Vine, and Branham, and the arguments of Newman, Dana, and Fager had bolstered the view that large starfish aggregations were natural occurrences doing little harm that nature itself could not take care of. If *Acanthaster* was actually undergoing a "population explosion" throughout the Indo-Pacific, the case would have to be strengthened. Chesher and Endean gathered further testimony from both scientists and nonscientists—divers, fishers, and fisheries departments throughout the Indo-Pacific—to argue that the population explosions were unprecedented and caused by human activity.

Divers exploring on the east coast of Africa between 1958 and 1969 reported that they had rarely seen *Acanthaster*. In 1969, when visiting several islands in the Indian Ocean, they saw "vast areas of devastated coral and large populations of *Acanthaster*." [47] They also noted that a passageway through the reef had been blasted on the leeward side of one of the islands to allow small boats through the coral to load phosphates mined on the island. In Fiji, new crown-of-thorns surveys were published in 1971 indicating that there, as in Western Samoa, Guam, and Australia, "man's activities are primarily responsible for the present outbreaks of *A. planci*." [48] Rather than single out any one human activity, some suggested that the outbreaks might be an overall effect of many destructive practices such as shell collecting—not only of tritons, but also of giant clams and pearl shells that might feed on the starfish larvae—handling and removal of coral and rock, blasting and dredging, and overfishing.

In 1972, Japanese researchers, who had visited the University of Guam, told of outbreaks in the Ryukyu Islands beginning in the late 1950s. The heaviest infestation occurred in the central part of the west coast of Okinawa starting in 1969.[49] Robert Jones and his collaborators at the University of Guam conducted a starfish survey in Taiwan, where they found large populations. From June 1970 to May 1971, the University of Guam resurveyed the reefs of Guam and twelve of the Trust Territory islands previously investigated by the Westinghouse teams. Truk, Pohnpei, and Guam had the most heavily infested reefs; they were also the most heavily populated by humans.[50] Increases in the number of starfish were reported by local people from Ulithi, Yap, and Kusaie in the Caroline Islands.[51]

John Randall, from the Hawaii Institute for Marine Biology, and Dennis Devaney, of the Bernice P. Bishop Museum in Oahu, conducted a six-month expedition in southeast Oceania to survey for *Acanthaster*. Supported by grants from the Bishop Museum, the National Geographic Society, the Oceanic Foundation, and the Sea Grant Program of the University of Hawaii, they visited twenty-five islands.[52] There was no question in their minds that the outbreaks were real and unprecedented. They were "massive invasions." Of all the islands they visited, only Tahiti and Rarotonga had large numbers of the starfish. They were by far the most populous islands of that part of the Pacific.[53]

Gathering such reports in 1973, Chesher and Endean argued that they clearly showed that crown-of-thorns infestations had occurred in recent years in widely separated areas throughout the tropical Indo-West Pacific region.[54] They offered a chronological account of this occurrence. The first major infestations were reported in 1957 in the Ryukyu Islands. In the early 1960s, certain reefs of the Great Barrier

Reef became infested, after which the number of infested reefs in the Great Barrier Reef increased markedly. In the Carolines, an area of the reef at Truk was reported to be infested in 1963; infestations of the reef at Tinian in the Marianas began in 1964 or 1965. By 1966 or 1967, several other reefs of the Carolines, Marianas, and Marshalls had become infested, and possibly certain reefs of Malaysia and New Britain at about the same time. Guam's infestation began about December 1966. In the late 1960s, infestations were observed in the Fiji islands, in the Philippines, Solomons, Western Samoa, the Cook Islands, the Society Islands, the Hawaiian Islands, and on reefs off Ceylon (Sri Lanka), and Taiwan.

There was no evidence of an overall decline in starfish numbers in any of the infested regions. Starfish numbers had decreased only on reefs where the bulk of hard coral cover had been killed or control measures successfully applied. In the Great Barrier Reef, the Marianas, the Carolines, and Fiji, areas devastated by the starfish had only increased. Chesher and Endean denounced the claims that the novelty of the infestations was merely due to the attention given them. The plagues made the publicity; the publicity did not make the plagues. They further debunked the evidence of Dana, who had enlisted statements from scientists forty years earlier to argue that *Acanthaster* was common in the Philippines and Palau before the Second World War. The author of the report about the Philippines personally told Endean that *Acanthaster* was "never known as a pest" there. The actual identity of the starfish said to be "very common" in Palau before the Second World War was questionable since it was reportedly found on "rocky and sandy" substrata.[55]

A set of three central arguments supported the belief that the infestations were caused by human activity: (1) The geographical issue—most of them occurred near centers of considerable human populations. (2) The novelty of the plagues—supported by lack of previous reports of such extensive outbreaks from scientists and indigenous people and estimates that some of the coral killed represented hundreds of years of continuous development. (3) The correlative timing of the plagues— the relatively brief time frame during which the infestations occurred in many different areas throughout the Indo-Pacific. For many biologists, it was asking too much to believe that they were natural cycles that coincidentally took place at many widely separated reefs at essentially the same time.[56] Endean and Chesher refrained from discussing the possible anthropogenic causes. Their favored theories—the triton, and dredging and blasting—had been well scrutinized by scientists on both sides of the debate. The possibility still remained that pollution from runoff, DDT, and other chemicals was the cause.[57]

DDT

In 1972, the same year in which Nierenberg declared the controversy closed and a nonproblem, John Randall published his paper in Hawaii: "Chemical Pollution in the Sea and the Crown-of-Thorns Starfish." [58] Randall was recognized as the world's premier coral-reef-fish scientist; he had also participated in the Westinghouse Survey in Micronesia and Hawaii. He noted Pearson and Endean's suggestion in 1969 that pollution from the mainland (agricultural chemicals, sewage, and effluent from sugar mills) may have led to the infestations. They had cited a period of heavy flooding and runoff in 1967 that produced discolored water as far as 50 kilometers (30 miles) offshore in regions later infested with starfish. There was still another significant clue, important for all subsequent theories. It came from new surveys in Micronesia led by Robert Jones. Reefs of low-lying atolls generally did not have as many starfish as those of high islands where more runoff would occur.[59]

Pesticide use, both for agricultural purposes and in homes, was "moderate to heavy" in well-populated areas with starfish infestations: Tahiti, Rarotonga, Guam, Upolu, and Fiji. In Tahiti, uncontrolled spraying with insecticides (especially dieldrin) had resulted in localized fish kills in lagoons. There had been heavy use of chlorinated hydrocarbons (first DDT and later lindane and dieldrin) in the Cook Islands. In Rarotonga, when people discovered that fish could easily be killed with small amounts of lindane and dieldrin, they promptly began to use them in streams, pools, and in lagoons. In Fiji, imports of pesticides and disinfectants (grouped for tariff purposes) increased from 26,358 pounds in 1960 to 99,205 pounds in 1968.[60]

Of all the pesticides, DDT was the logical first suspect for the ecological imbalance leading to starfish population explosion. During the period 1962 to 1966, the average production of DDT increased nearly 8% per year. In 1966, more than 625,000 tons were produced globally.[61] Its use had declined in the United States and Europe because of insect resistance and public indignation over its effect on birds and aquatic life. Nevertheless, huge amounts were still shipped for use in other countries.[62] There was a flurry of scientific studies in the 1960s about its persistence in the environment. As a component of dust, it is wind-borne for great distances. Studies by the Fish and Wildlife Service of the Department of the Interior in 1963 indicated that more than 1000 pounds of DDT were transported across the Atlantic from Europe and Africa an-

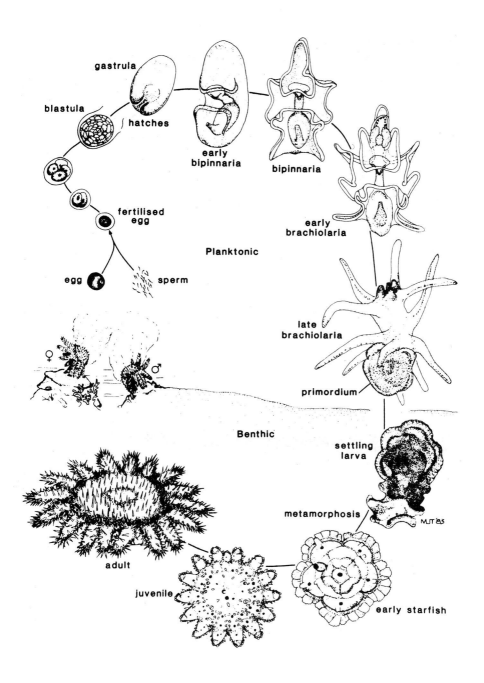

The life cycle of the crown-of-thorns starfish. Drawing not to scale. (From Peter Moran, *Crown-of-Thorns Starfish; Questions and Answers*, AIMS, 1988.)

Taken in American Samoa, this photograph shows the aggregating behavior of feeding crown-of-thorns on progressive bands of (1) uneaten green coral, (2) a band of freshly eaten white skeleton, (3) a shady band of four- or five-day old algal growth, and (4) progressively darker bands of algae. (Photograph by Charles Birkeland.)

A wall of marauding crown-of-thorns moving down the coast of American Samoa. The white is the skeleton of freshly eaten *Acropora*. Notice how the crescent-shaped wall of starfish wraps around the top of the photograph. The wall is heading to the left. (Photograph by Charles Birkeland.)

Close-up of the crown-of-thorns showing spines and many arms. The patch of coral recently eaten by the starfish (left) is called a feeding scar. (From Peter Moran, *Crown-of-Thorns Starfish: Questions and Answers*, AIMS, 1988.)

Single feeding crown-of-thorns in shallow water on the Great Barrier Reef. (Photograph by Peter Moran.)

A giant triton, *Charonia tritonis*, feeding on an adult crown-of-thorns starfish. (From Peter Moran, *Crown-of-Thorns Starfish: Questions and Answers*, AIMS, 1988.)

This starfish is pried away from the coral to show its membranous stomach. (From Peter Moran, *Crown-of Thorns Starfish: Questions and Answers*, AIMS, 1988.)

Crown-of-thorns off the coast of Panamá. (Photograph by Peter Glynn.)

Bleached coral close-up, Uva Island Reef, 22 March 1983. (Photograph by Peter Glynn.)

The beginning of an outbreak of crown-of-thorns on the Great Barrier Reef. Note the numerous white skeleton remains of diverse coral species. (From Peter Moran, *Crown-of-Thorns Starfish: Questions and Answers*, AIMS, 1988.)

Mixed bleached and dead coral, Indo-Pacific puffer fish. (Photograph by Peter Glynn.)

Painted shrimp (*Hymenocera*) on the back of the crown-of-thorns starfish. The relationship between the shrimp and the coral it protects from the starfish is called "guarded mutualism." (Photograph by Peter Glynn.)

Diadema in sea grass in St Croix in 1974. Nocturnal forays of Diadema from patch reefs into surrounding sea grass beds create "halos" of grazed sea grass. Diadema also prevents fleshy algae from overgrowing and smothering the coral. The mass mortality of these sea urchins in 1983 helped to show marine ecologists that the Caribbean functioned as a large ecosystem. (Photograph by John Ogden.)

nually. Ocean currents move it widely in the sea, and it was carried in the bodies of migratory fishes and birds. There were reports in *Nature* announcing residues in penguins and seals in the Antarctic.[63] There were estimates, in the early 1970s, that 1.5 million tons of DDT were still adrift in the biosphere.[64]

DDT was developed during the Second World War, its use was widespread in the 1950s, and, Randall argued, it may have taken ten years to attain the levels of contamination necessary to cause the ecological disaster. He suspected that the effect of pesticides would not be on the predators of adult *Acanthaster* but on those of the starfish larvae. During their approximate month of life in the open sea, the starfish larvae feed on phytoplankton (unicellular algae) that were known to build up concentrations of DDT that greatly reduced their photosynthesis.[65] Each link in the food chain from phytoplankton to larval starfish and the predators of those larvae would successively acquire higher concentrations of pesticides until one or more of the starfish predators would acquire concentrations that were lethal or would reduce its reproduction.[66]

To test this hypothesis, one needed more field data, particularly in areas remote from centers of human population and agriculture. If pesticide residues in the sea were responsible, one should also expect gross fluctuations in the populations of other marine organisms. The crown-of-thorns was noticed because of its large size and disastrous effect on coral. But it raised the question: how many other excessive fluctuations in the abundance of marine life had gone unnoticed?[67]

DECLINE IN BIODIVERSITY

Many marine ecologists feared an increase and spread of *Acanthaster* infestations due to existing and planned activities of humans along tropical coastal areas. Even the prevalent view that the starfish played an important role in maintaining coral diversity was refuted—by Peter Glynn at the Smithsonian Tropical Research Institute (STRI). Brought up in Coronado, south of San Diego, California, Glynn had become interested in marine biology as a child collecting shells on the beach, and diving since the age of twelve.[68] He was president of a skin diving club during the Second World War. The Underwater Demolition Team, which the United States used in the Pacific to fight the Japanese, was located in Coronado. As Glynn recalled, they wanted to reach out and get into the community to help young kids, so they adopted him as their "mascot" around 1945 and put him through their training exercises.[69]

Glynn completed his Ph.D. in 1963 at Stanford University's Hopkins Marine Station, Pacific Grove. After working at the University of Puerto Rico, he moved to STRI in 1967, where he remained until 1983. Today, he is Professor at the Rosentiel School of Marine and Atmospheric Science, University of Miami. Glynn's research encompassed coral-reef community structure, species interactions, disturbance responses, and recovery. He is perhaps best known for his pioneering studies in the 1980s of widespread coral bleaching and its relations to ocean temperature increases and enhanced El Niños. In the 1970s, he focused on the crown-of-thorns. As he explained:

> The crown-of-thorns was extremely interesting because, like any physiologist or experimentalist a lot of times you don't understand systems until you hit them over the head with something like a mutation—or you stress them somehow. So I thought the crown-of-thorns was an ideal predator-prey system. . . . we could really see how it was affecting the dynamics of a coral assemblage. So that was the whole approach I took in looking at the crown-of-thorns.[70]

Coral reefs in the eastern Pacific are not as plentiful or as lush and diverse as those in the western Pacific. At STRI in the 1970s, Glynn was interested in what effect *Acanthaster* had on coral-reef growth there, and he wanted to resolve the highly speculative issue of the amount of coral predation a reef community could sustain.[71] No crown-of-thorns outbreaks were reported in the eastern Pacific, and Glynn suspected this was because of adequate starfish-predators at every stage of its life cycle to keep normal populations in check. There were also few funds to study the crown-of-thorns in the United States at the time. But, founded in the 1920s, and expanded in the 1950s and 1960s, STRI was the premier American institution of its kind, with twenty-five full-time researchers and an extensive library. One could apply for internal research grants supplied by the Smithsonian Institution through awards from the U.S. Congress. STRI also received funds from the Tupper Foundation. Then as now reef studies were funded chiefly because of their importance to fisheries and tourism. They were also funded by the National Institutes of Health and the National Science Foundation on the basis that new chemical compounds could be found in some species that might be useful for cancer research. For example, certain sponges harbor acromycin, an important antibiotic, second only to penicillin. Glynn justified his studies of the crown-of-thorns in accordance with these interests:

> Because of the paucity of coral reefs in the eastern Pacific region and the various benefits which they offer (e.g., unique fishing-grounds, a source

of novel organic compounds for use in pharmaceutical research, and underwater parks of value to tourism), it is desirable to evaluate the seriousness of this problem in the Americas.[72]

That the crown-of-thorns was a keystone species that would increase coral species diversity by preferentially feeding on fast-growing, numerically dominant coral, thereby opening up spaces for relatively rare coral species had been suggested by Porter and presented by others including Branham, Newman, and Dana as an argument against eradication programs.[73] This conclusion was intuitively sound and it appeared to follow well-documented cases involving keystone predators.[74] But, as Glynn saw it, the actual evidence was circumstantial and weak. He studied *Acanthaster* prey preferences in detail on the Pacific coast of Panamá.

Glynn agreed that the fast-growing genus *Pocillopora*, the dominant coral in the eastern Pacific, constituted a high proportion of the corals eaten by *Acanthaster*. But this did not necessarily mean that the starfish preferred *Pocillopora*. One had to consider the relative availability of all potential coral prey. Perhaps *Acanthaster*'s apparent preference for *Pocillopora* was due only to the latter's relative abundance. In fact, Glynn's detailed studies indicated that the starfish fed proportionally more on relatively rare corals than on *Pocillopora*.[75] *Acanthaster* was exploiting its food resource in "a coarse-grained fashion to the detriment of the less common species." [76] Far from increasing coral diversity, he concluded in 1974, the starfish's "prey preferences for the less common corals would work in the opposite direction, i.e., would tend to depress community species diversity." [77] During the 1970s and 1980s, Glynn turned to study one of the crown-of-thorns potential predators, the painted shrimp.

GUARDED MUTUALISM

When, in 1970, the painted shrimp was first observed in captivity feeding on the adult starfish, it was widely speculated that it might be an important predator for keeping the starfish populations in check. However, in subsequent field studies on the east African coast, Wolfgang Wickler noted that it fed largely on other smaller and more sedentary seastars.[78] This view seemed to be confirmed by studies in the Red Sea by Rupert Ormond and Andrew Campbell, who noted that the shrimp preferred shallow lagoons on reefs where *Acanthaster* was uncommon.[79] Glynn's observations in Panamá were in direct contradiction: the shrimp and the starfish did cohabit large areas of the reef and the shrimp commonly preyed on *Acanthaster* and interfered with its feeding activities.[80]

Both the starfish and the shrimp were present along the seaward reef slope that Glynn studied.[81] The shrimp commonly took shelter in the branches of live *Pocillopora*. When the starfish attacked the coral, it attacked the starfish. Glynn's study of two pairs of shrimp is illustrative. When first observed, *Acanthaster* had extruded its stomach over a portion of a coral colony; one pair of shrimps were on the seastar's back, attempting to pierce the dermis. After two hours, when the starfish finally moved off the coral, only a 3-cm^2 area of the coral colony was killed. Over the next fifteen hours, the starfish tried to assume a feeding posture on several different coral species, but the shrimp again interfered. "The second large pair of shrimp climbed on top of *Acanthaster*, successfully pierced the dermis (by amputating 10-cm and 3-cm terminal segments) and removed tissues from the gonads and hepatic caeca. These shrimp also interfered repeatedly with the feeding posture and stomach eversion of *Acanthaster*." [82] Glynn made similar observations in a large tank: "A typical reaction for *Acanthaster* that had a previous encounter with *Hymenocera* was to climb the tank wall and remain stationary for several days, even when the shrimp was absent." [83] So for Glynn, it was not suprising that *Acanthaster* was lacking in regions where the shrimp were present in the Red Sea—the shrimp kept the starfish away.[84] In further studies in the Gulf of Oman, he observed certain crabs and snapping shrimp protecting *Pocillopora* as well.[85]

It was clear to Glynn that *Acanthaster* predation could drive coral species diversity to zero in some locations. Although there were no plagues of *Acanthaster* in the eastern Pacific, nothing seemed to be stopping them from reaching that region. Tropical starfish larvae were known to be capable of long-distance travel. There was still another reason for Glynn's concern about the crown-of-thorns: the proposal to excavate a sea-level canal through the Central American isthmus. *Acanthaster* was not present in the lush coral reefs of the Caribbean Sea. But even this, he warned, could be changed. Great fear over the canal proposal was widespread among biologists at STRI and elsewhere during the late 1960s and early 1970s.

8

OCEANS APART

The fact that *Acanthaster* is separated from the Caribbean Sea by a freshwater barrier (the Panamá canal) which could be altered by Man in the near future, demands the closest scrutiny.

Peter Glynn

I consider it my duty as a human being and a lover of nature, when I see a potential disaster to nature, to do something about it. Call it politics, or no, but I don't call it politics.

Ernst Mayr

SWORD TO PLOWSHARE

The history of the Panamá Canal is inexorably linked with international politics, economics, and military concerns.[1] The Spaniards of the sixteenth century planned a canal and in the seventeenth century fairly large ships were moved across the isthmus on rollers.[2] Not until the end of the nineteenth century was serious construction started in what was then the Colombian province of Panamá. In France, *Campaigne Universelle du Canal de Panamá* was formed. In 1878, it obtained rights from the Republic of Colombia to build a canal across the isthmus where the continental divide dips to one of its lowest points. It was led by the French engineer Ferdinand de Lesseps, who had been successful in building the Suez Canal, which opened nine years earlier. Digging the Panamá Canal began in 1881, but the attempt failed. More than 22,000 workers died, mostly from yellow fever and malaria; the company went bankrupt, and the attempt was given up in 1889.

In 1902, the United States bought out the French interests and began talks with Colombia for the rights to build a canal, but Colombia was unwilling. Because of this intransigence, the United States gave its tacit support to a rebellion in Panamá in 1903. President Theodore Roosevelt seized the opportunity to send a cruiser to Panamá, supporting its secession from Colombia while quickly gaining agreement from the new Republic of Panamá to build the canal. Walter Reed and his colleagues identified the yellow fever vector, and construction under the direction of John Stevens and George Goethals began in 1904 and continued for ten years, employing as many as 40,000 men at one time. It involved the removal of more than 153 million cubic meters (200 million cubic yards) of soil and rock. The canal opened in 1914. Often recognized as "the greatest engineering feat of the modern age," it effectively cut the sailing distance from New York to San Francisco by about 12,000 kilometers (7000 miles).

The treaty of 1903 called for the establishment of "the Canal Zone," a strip of land extending about 8 kilometers (5 miles) on either side, which would be controlled by the United States. In return, the United States agreed to pay Panamá $10 million and an annual rent of $250,000. The treaty was modified in 1936 when the United States relinquished its eminent domain within the Canal Zone and raised its annuity to $430,000. However, Panamanians believed their sovereignty was continually threatened by American military occupation for canal defense and felt deprived of their birthright. The canal was heavily fortified, with brigade posts near the locks. Both the army and navy had aerial forces at the canal; a base for submarines was maintained at the Atlantic end. In 1964, following riots between Panamanian students and U.S. citizens living in the Canal Zone, President Johnson appointed a negotiating team to draft a new set of treaties that would include granting joint jurisdiction (with a Panamanian minority) over the Canal Zone, increasing the royalty payments to Panamá, and giving the United States a twenty-year option to build a sea-level canal. Panamanians rejected the entire package as politically unacceptable. At the same time, strong opposition existed in the American Senate to relinquish American "sovereignty" over the canal zone.[3]

In 1965, President Johnson asked Congress to establish a five-member Canal Study Commission to lay the groundwork for a sea-level canal project. The Atlantic–Pacific Interoceanic Canal Study Commission was composed of a diplomat, a lawyer, a university president, a former army engineer, and a civilian engineer.[4] Congress authorized $24 million for a feasibility study and the commission's task was to recommend a location for a second canal, study the scope of the anticipated negotiations with the country involved, recommend an excavation technique, assess costs and means of support, and consider a defense system for the canal. It was to report back to the President by December 1, 1970.

A sea-level canal was believed to have several merits over a lock-type canal. By 1970, traffic through the canal had increased dramatically. About 13,000 ships passed through annually, and the canal was projected to reach its capacity of 19,000 ships per year by 1985, thus cramping U.S. and world trade. Ships had also gotten larger. There were estimates that some 1400 ships, huge tankers and bulk carriers, plying the seas could not pass through the canal because of draft and beam limitations. The operation and maintenance of a sea-level canal would be easier and cheaper; fewer personnel would be required, and repair and overhaul of locks and equipment eliminated.[5] From a security point of view, a sea-level canal was less vulnerable to being damaged and put out of operation. In the event of an attack, the vital water supply from Gatun Lake, which feeds the locks, would not be lost. A sea-level canal would also be able to transit large aircraft carriers that could not fit through the existing locks.[6]

The idea of constructing a sea-level canal through Panamá had been seriously contemplated in the United States in 1904, but had been abandoned in favor of the lock-type. Construction to improve the canal with larger locks and an extra set for the purpose of defense had begun in 1939, but other priorities intervened when the United States entered the Second World War and work was abandoned in May 1942, after about $75 million had been spent. These excavated locks still lie unfinished at the Pacific end of the canal.[7]

With the development of nuclear energy during the war, the construction of a sea-level canal was considered one of the ways of putting this new source of energy to "peacetime use." [8] Indeed, this peacetime use was one of the key factors that had prompted President Johnson to appoint the commission with the hope, as one commentator put it, that "if this could be done in some spectacular fashion it might capture the public fancy." [9] The Army Corps of Engineers, who wanted to accomplish the task, estimated that a total of 170 megatons of nuclear charges would do the job. Battelle Memorial Institute had a contract with the Atomic Energy Commission to examine the effects of radiation fallout on flora and fauna, as well as exposure to human populations in the study area.[10] Founded during the Great Depression, with laboratories in Columbus, Ohio, to deliver technology-based solutions to commercial and industrial problems, Battelle later emerged as "the world's largest, independent science and technology institute." Its Environmental Systems and Technology Division assists global organizations, companies, and government agencies to develop strategic environmental, health, and safety management systems.

Even if atomic charges were not used, one aspect that received little attention from the Canal Study Commission was the ecological effects of a sea-level canal on ocean life. The Atlantic and Pacific marine biotas were thought to have been separated for three to five million years when the land bridge joining South America and Central America was formed. The estimated 8000 marine species in the Pacific and 7000 in the Caribbean had markedly diverged from whatever similarities they originally held in common.[11] A sea-level canal would constitute an unobtrusive two-way transport system for dispersal of free-swimming, shallow-water stenohaline marine fishes and at least a one-way system for planktonic stages.[12] The sea-level canal exploded into controversy. The Smithsonian Tropical Research Institute (STRI) first brought the matter to the attention of the Commission in April 1966.[13]

THE GREATEST EXPERIMENT IN HISTORY

Ira Rubinoff, then assistant director for Marine Biology at STRI, was one of the first to emphasize the potential effects on oceans. As a student of the Harvard evolutionist Ernst Mayr, Rubinoff had first gone to Panamá in the early 1960s to investigate the evolutionary divergence of fish species on both sides of the Panamanian isthmus. Mayr himself had conducted a study of speciation in echinoids (sea urchins) that showed that when the isthmus rose up millions of years ago, it had separated a continuous population between the Atlantic and the Pacific, leading to the evolution of six different genera of sea urchins.[14] With Rubinoff's interest in fishes, Mayr sent him to Panamá to carry out similar studies. Mayr had also worked on studies of birds with the director of STRI, Martin Moynihan. Moynihan was a well-known naturalist, whose work on the evolution and communication of New World monkeys, marine animals, and birds was highly celebrated. He was also noted as an outstanding adminstrator, and for the feat of turning the research station on Barro Colorado Island, established in the 1920s, into the Smithsonian Tropical Research Institute he was awarded the Smithsonian's Joseph Henry Medal.[15]

After completing his Ph.D. in 1964, Rubinoff returned to Panamá as Moynihan's assistant to establish marine laboratories on both sides of the isthmus. His own talents at logistics, administration, and skillful di-

plomacy soon became obvious. Moynihan resigned from the director-
ship to return to research in 1973. Rubinoff was appointed director of
STRI the following year, a position he holds today, maintaining STRI as
United States' premier tropical research institute. He had the ability to
get along with the American military, "the tail that wagged the whole
dog," as Mayr put it.[16] He was also able to work with the local government as well as the Smithsonian administration in Washington.

In 1968, in an oft-cited article in *Science*, Rubinoff described the inevitable making of a sea-level canal as the "greatest experiment in man's
history." It was one that came along only once in about five million
years and now had to be exploited "for advancing our scientific understanding of evolutionary and ecological processes."[17] Rubinoff warned
of the possible consequences: inferior hybrids might be formed from
mixing species, which would lead to the extinction of both species, or
the replacement or extinction of one species by another. Ecologists by
necessity generally studied short-lived and localized perturbations, such
as the burning of forests, the felling of trees by wind, or the effect of
grazers on their food. Long-term and global events, such as species extinction, biotic effects of climate change, and faunal upheavals were
normally the domain of paleontologists. The large gap between ecological and paleontological temporal and spatial scales was unstudied because global effects rarely occurred naturally. But, in the case of building
a sea-level canal, "global" effects would be produced "experimentally."
Rubinoff called for the establishment of a multidisciplinary scientific
control commission with broad powers of approving and disapproving
major alterations to the marine or terrestrial environments any place
where the U.S. government or private contractors might be active.[18]

A range of opinions existed among biologists in response to the proposed sea-level canal. At the University of South Florida, zoologist John
Briggs estimated that between 1000 and 5000 species of marine animals could possibly perish due to competition.[19] He reprimanded
Rubinoff for assuming that a sea-level canal would be constructed and
for regarding its advent as an opportunity to conduct a one-in-five mil-
lion-years experiment.[20] Briggs in turn was scorned by other biologists
for being "alarmist" and "dogmatic," and for his use of the word "disaster" when describing the potential outcome.[21]

The risks to marine life were also trivialized by the President's Canal
Study Commission. And no one minimized them more than did its Executive Director, John P. Sheffey, a retired army colonel. Research on
potential environmental changes would be "nice to have, but not very
important." He added, "We can't be certain of the biological implica-

tions, until after the canal is built anyway—regardless of how much research is done now." But in his view, the "possibility of any serious
disruptions to nature are very remote, and the potential threat to biota
is so insignificant that it doesn't merit spending a lot of money on it." [22]

After all, he argued, there had already been large transfers of marine
life through the existing lock canal without environmental upsets and
this had gone on for more than half a century. The existing canal, with
its intervening 40 miles of fresh water, Gatun Lake, served as an effective barrier to some species of marine life. But there were swimming and
drifting biota that survived the journey through the fresh water lake and
passed readily through the locks. Barnacles and similar clinging organisms also passed in both directions every day on the hulls of ships. Most
important, there was daily transfer of marine life in the saltwater in
ship's ballast tanks.[23] Lightly loaded or empty ships approaching the canal are frequently required to take on ballast water before entering the
locks. This enabled them to deepen their drafts and made them easier to
handle while in restricted canal channels. On leaving the canal a few
hours later at the opposite ocean, this ballast water is usually discharged
to lighten the ships and save fuel on the remainder of the trip.

Therefore, Sheffey argued, all the small swimming and drifting marine life in these thousands of tanks of seawater, discharged year-in and
year-out since 1914, had made the trip across the isthmus in both directions with no discernible effects.[24] Sheffey's critics argued that it was absurd to compare ballast-water dumping to free swimming from one
ocean to the other.[25] But it was obvious to him that those environmen-
tally concerned biologists had simply adopted "a policy of taking an
alarmist view to attract attention, and they tacitly admit it." [26]

The Nobel laureate geneticist Joshua Lederberg would have to be included among those Sheffey deemed "alarmist." In the *Washington Post*,
in February 1969, he discussed the sea-level canal proposal while warning of accelerated global environmental modification that accompanied
the rapidly increasing human population.[29] "This generation is the first
to have seen mega-experiments that had global consequences far too
rapidly to hope to measure all the consequences." Lederberg was referring to nuclear bomb tests with hazards of global radioactive fallout and
pesticides like DDT. Certainly, scientists' judgments about such ecological hazards usually fell short of the rigorous proof to which they cus-
tomarily appeal. But, for "many mega-experiments," Lederberg argued,
that was an unachievable standard. "The planet could be committed to
the ash heap," he warned, "before such a fate was proven by the standards of laboratory experimentation." He prophesied that the conse-

quences of a sea-level canal would be dramatic and irreversible: some species would be wiped out, others would find new homes and flourish; still others might hybridize to give sterile offspring, possibly endangering both the Pacific and Atlantic varieties. There would be serious economic consequences on fisheries as well. "And this is the very industry in which we repose great hopes for improving the world's food supply." Lederberg was encouraged when newly elected President Richard Nixon was urged by his preinaugural task force on resources and environment to name a "Special Assistant for Environmental Affairs" to work closely with him.

There were risks of still other irreversible effects. Nothing provided a more dramatic illustration of obvious risk than the crown-of-thorns. Here was a poisonous spiny species — not found in the Caribbean — with devastating effects on coral reefs and whose potential harm to coral reefs in the Caribbean was surely far too serious for anyone to risk. The only other organism comparable was the poisonous yellow-bellied sea snake, also not found in the Caribbean. As John Ogden recalled, "There was all this concern about the sea-level canal that was fastened onto major organisms; one of them was the sea snake —the deadly poisonous sea snake—and the other was *Acanthaster*." [28] Glynn repeatedly emphasized the same issue: "The unexpectedly broad tolerance limits of adult *Acanthaster* and its coral prey should warn us of the risk involved in altering the present freshwater canal barrier in any way that would enhance transisthmian migrations of marine forms." [29] Robert Johannes did the same: "Crown-of-thorns starfish are not presently found in the Atlantic. Construction of a sea-level canal through central America could facilitate their spread, via their planktonic larvae, into the Caribbean, where their potential for damage to the reefs is frightening to contemplate." [30]

CERIC

Environmental concerns were sufficiently vocal that, in February 1969, the Canal Commission asked the National Academy of Science to appoint a committee of biologists to examine the ecological issues with special reference to the marine environment. Rubinoff asked Mayr to chair the committee. Mayr was an active researcher, one of the foremost naturalists and evolutionists of our time. He was involved in developing the Museum of Comparative Zoology at Harvard in 1969. As he put it, "I wouldn't have dropped all that and got involved in the sea-level canal, if I didn't think that there was a potential there for disastrous effects on the environment." [31] Chaired by Mayr, the Academy's committee, called the "Committee of Ecological Research for the Interoceanic Canal" (CERIC), was made up of a distinguished group of ecologists, naturalists, and marine scientists: Máximo Cerame-Vivas, University of Puerto Rico; David Challinor, director of International Affairs, the Smithsonian Institution; Joseph Connell, University of California, Santa Barbara; Ivan Goodbody, University of the West Indies, Jamaica; William Newman, Scripps Institution of Oceanography; E. O. Wilson, Harvard University; C. S. Ladd Prosser, University of Illinois; Howard Sanders, Woods Hole Oceanographic Institution; and Donald Wohlschlag, University of Texas.

CERIC was to assess the ecological impact of a sea-level canal and recommend ways of minimizing the damage that might occur. It was also charged with the responsibility of outlining a program of research to be done in the period before, during, and after the canal construction. The actual need for a sea-level canal, and the wisdom of constructing it, were explicitly excluded from the committee's task.[32] Nonetheless, its assessment of the consequences and how to minimize serious ecological affects became the source of heated controversy.

CERIC decided that some sort of barrier to marine life had to be incorporated into the design of any sea-level canal. Otherwise, the yellow-bellied sea snake and the crown-of-thorns starfish could easily move through the canal, which would also provide an optimal habitat for certain large Pacific sharks. The tourist trade would be seriously effected. So too would commercial fishing—some species, including certain shrimp, could be replaced by economically less valuable ones. Parasites and pathogens could easily pass from one ocean to another, destroying organisms that lacked natural resistance to them.[33]

Assessing the risks was hardly a guessing game. CERIC pointed to previous canal projects where barriers had been eliminated that led to economic disaster for certain fishing industries. The invasion of the Great Lakes by the sea lamprey, a predatory fish-like creature found in the North Atlantic was a striking example. For thousands of years, the sea lamprey was barred from the inner Great Lakes by Niagara Falls, but a system of man-made canals allowed the lamprey to penetrate the inner lakes where it fed ravenously on the valuable lake trout and other fish. It was an economic nightmare for the fishing industry. The annual catch of trout in Lake Huron and Lake Michigan fell from 8.6 million pounds to 26,000 pounds in a period of only ten years. Effective control was achieved only after years of research and a costly management program.[34] Transmigration and colonization of marine plants and animals between the Red Sea and the Mediterranean had also occurred with the making of the Suez Canal a hundred years earlier. Large-scale population changes resulted, with significant economic impact. For example, a certain valuable species of sardine found in the eastern Mediterranean seemed to have been considerably affected by competition from a less desirable species that invaded through the Suez Canal from the Red Sea.[35]

CERIC considered such physical barriers as electronic weirs and ultrasonic screens, or treating the waters contained within the canal in some way. But making an antibiotic barrier would be no simple matter. There is a marked difference in sea-level and tidal amplitude between the two oceans. The Pacific side has a tidal range of 21 feet (6.3 meters) and is on the average 0.7 feet (21 cm) higher than the Caribbean, which has a tidal range of but a few feet. There would be a major flow of water from one ocean to the other through a sea-level canal. Tidal barriers would have to be arranged in such a way as to keep the net flow to a minimum so that the contained waters could be treated and maintained appropriately. If the flow could be kept to a minimum, CERIC suggested that a freshwater barrier might be investigated by engineers. This might be combined with thermal barriers: virtually all marine organisms would be killed by a residence time of forty-eight hours in less than 5% seawater and temperatures of 45°C.[36]

The Academy's committee recommended that extensive research on the physiology and natural history of the biota and the population dynamics of Panamanian fisheries should begin about ten years before the opening of a sea-level canal. Without background information there would be serious impediments to understanding and explaining changes that would occur after a canal was completed. Field studies needed to be conducted on nearshore communities, such as mangrove swamps, mudflats, soft bottoms, and coral reefs. Extensive oceanic work would also require major funding, space, ships, and adminstration. CERIC further recommended that a "Commission on the Ecology of the Interoceanic Canal" be established with a governing board of North, Central, and South American scientists.[37]

FALLOUT

Reverberations occurred immediately after CERIC's report went to the Canal Study Commission in April 1970. Two letters, one from the Canal Commission's director, John Sheffey, and another from its engineering agent, Brigadier General R. H. Groves, were sent to the executive officer of the National Academy of Science and circulated among members of CERIC.[38] Sheffey accused the Academy's committee of taking an extreme "alarmist viewpoint." He insisted that there was no need to consider special barriers to prevent free interchange between the tropical American biotas because there was simply little or no ecological risk. He asserted that the crown-of-thorns starfish would have already established itself in the Caribbean if conditions there were favorable. He also implied that he could rally experts of his own in support of his view.

Several CERIC members responded, notably William Newman. Far from CERIC's being alarmist, Newman retorted, it was Sheffey who was taking an extreme stand. He doubted that Sheffey could assemble qualified experts of his own to counter CERIC's assessment of the risks. He was confident that no knowledgeable person would allow himself to be held responsible for the undesired consequences that could result from the introduction of such organisms as the crown-of-thorns, or the yellow-bellied sea snake. Sheffey's statement that the crown-of-thorns would have already established themselves in the Caribbean only displayed "an ignorance of the facts" and of "elementary ecology." [39] From a purely ecological point of view, Newman argued, no canal should be built. Desirability could only be couched in terms of "commerce and defense," which were neither within the purview nor competence of CERIC to assess.[40]

Newman later recalled that the potential introduction of the sea snake to the Caribbean was really of no great concern, in his opinion. Although some species are deadly, the yellow-bellied sea snake, the only one in the Eastern Pacific, was only mildly poisonous and tended to swim in the open water rather than sitting around the bottom where one might touch it. Nonetheless, Rubinoff and Kropach demonstrated that its effect on other species would be dramatic.[41] As Newman recalled, the sea-snake "was just the tip of the iceberg, so to speak, and since it was something administrators could relate to, it was waved about, as was *Acanthaster*." [42] As for the crown-of-thorns, it was a different matter: "No one in his right mind would allow its introduction into the Caribbean since, while I doubted that it would take, at the level of our knowledge at the time the risks would have been too great." [43]

In fact, Newman and Thomas Dana proposed to carry out a controlled, closely monitored introduction of the crown-of-thorns to a fenced Caribbean patch reef to see what would happen in 1971. To prevent the starfish from accidently getting established in the Caribbean they would introduce only males. They made no other predictions other than to note that it might demolish the patch reef, it could be controlled by predators, or it might languish and eventually die, as did several introductions of the Maine lobster to the west coast of the

United States. When the experiment was over, any remaining starfish were to be sacrificed. However, in the biopolitical context of the sea-level canal, their research grant proposal "drew blanks." [44] Later, Newman and Dana suggested another experiment on an established population of *Acanthaster* on the Pacific coast of Panamá. It was designed to test an hypothesis of Glynn's regarding the amount of coral predation a reef community could sustain. This research proposal was also rejected. With a passing note in *Science* in 1974, suggesting the experiment to be performed, Newman left the *Acanthaster* problem to others who lived in the tropics. [45]

EXPERTS OF THEIR OWN

The final report of the Canal Commission was submitted to President Nixon in the fall of 1970. It recommended that a sea-level canal be built about ten miles west of the present Canal at cost of about $2.88 billion. [46] The Canal Commission had spent $17.5 million, about 75% of its budget, on evaluating nuclear excavation routes. [47] However, the use of nuclear charges was not recommended. The radioactive fallout would be too dangerous. Conventional excavation techniques had to be used because "neither the technical feasibility nor the international acceptability" of nuclear excavation had been established." [48] The Commission devoted only four pages of its 109-page cover report to a chapter on "environmental considerations" and the thrust of its conclusions was that whatever ecological risk might exist was "acceptable." [49] The voices of concerned biologists, and CERIC's year-long study, were not heeded. However, the Canal Commission's report did contain a summary of Battelle Memorial Institute's conclusions about potential ecological effects.

The Battelle team represented the "experts" Sheffey said he could call on. They minimized the risks and endorsed his previous statements that certain forms of marine life had been passing through the existing canal for fifty years on the hulls of ships and in ballast water, with "no harmful results" identified. [50] There was "no firm evidence to support the prediction of massive migrations from one ocean to another followed by widespread competition and extinction of thousands of species." [51] They asserted that differences in environmental conditions on the two sides of the isthmus, coupled with prior occupancy of similar ecological niches by analogous species, would act as natural deterrents to the establishment of any species that might travel through the canal. They considered it "highly improbable that blue-water species like the sea snake and the crown-of-thorns starfish could get through the canal except under the most unusual circumstances." [52] They also found no evidence that commercial or sport fisheries would be affected. In assessing the risks, the Canal Commission adopted the same rhetorical model as the intergovernmental crown-of-thorns committee in Australia: if it cannot be proven, the threat is insignificant.

Finger-pointing, aiming blame at the National Academy of Science's committee for the Canal Commission's lack of regard for the ecological consequences, was not long coming.[53] Ernst Mayr told a science reporter in 1971, "We said that great danger would result from building a sea-level canal, though we can't prove it. But they turned it around and said that, since we can't prove it, the danger is minimal." [54] Moreover, CERIC had been given a restricted role from the outset since its deliberations were carried out on the assumption that a canal would be built. As Mayr explained, the Canal Commission had told his committee members, "'Look here, boys. That canal is going to be built no matter what you say.' Consequently, we decided the best thing to do was to make the canal as harmless as possible." [55]

Despite the Canal Commission's disregard for environmental risks, the sea-level canal plan was not followed. As Ogden put it, "cooler heads prevailed." Not for environmental reasons, but for political ones—Panamá wanted complete control. The billions of dollars in money and technology were all to be provided to Panamá with no strings attached. In 1977, a new treaty was signed that provided for Panamá to take control of the Canal Zone in 1979, and the canal itself by 2000.[56] But the idea of building a sea-level canal is still adrift.[57]

9

REMOTE CONTROL

Inevitably, there will always be those who disagree with an assessment of this
kind, particularly individuals whose views have been swayed on emotional
grounds. Emotional commitment to a cause may not, in itself, be wrong, but
it should follow rather than precede a dispassionate examination of all the
relevant evidence.

Crown-of-thorns Advisory Committee, 1975

The notion that the starfish outbreaks might be natural and relatively
harmless found new scientific support in the early 1970s. Special funds
for crown-of-thorns research were made available in Australia. Following
the recommendation of the joint government Committee of Inquiry, a
crown-of-thorns Advisory Committee was established to recommend
grants-in-aid.[1] Four of its members had been on the Committee of Inquiry.[2] Advertisements inviting applications were placed in all major Australian newspapers and posters were displayed at all Australian universities
and other research institutions. From 1971 to 1975, the Committee
awarded over $A440,000 to projects on reef ecology, starfish larval devel-
opment, coral formation, as well as monitoring and control measures.
Public seminars were held in Brisbane in 1972 and again in 1974 to en-
able grantees to meet and discuss their research. Interested members of
the public were invited to attend and participate in discussions.

In 1975, the Advisory Committee published a report on the results of
the research it supported.[3] Its report read as a biting response to public

outcries and criticisms of the Committee of Inquiry five years earlier. In the course of that previous committee's investigations, "No factual evidence had been presented . . . to support the extreme views on imminent reef destruction. On the contrary, many of the assertions made by proponents of these views were withdrawn or significantly qualified under cross-examination." [4] Furthermore, the results of new research had clearly "vindicated the assessment made by the Committee of Inquiry." The Advisory Committee organized the results of the most productive research into a coherent whole to argue that nothing unnatural was happening to the Great Barrier Reef. Indeed, it promoted a new theory for the increases in *Acanthaster* populations. It was based on laboratory experiments on temperature and salinity conditions for larval survival by John Lucas at James Cook University in Townsville, North Queensland.

Despite the research funds, the move to study the crown-of-thorns was not straightforward for many Australian scientists. Educated at the University of Western Australia in Perth, Lucas had specialized as a crustacean biologist, investigating spider crabs. After completing his Ph.D. in 1968 and moving to Townsville, he was determined to work on a Great Barrier Reef animal. But he soon discovered that this was not logistically possible.[5] There were no research vessels or reliable access to the Reef from Townsville. So he reverted to studying inshore crabs and found a great diversity of them in mangrove habitats. He would have continued this line of research except that Julie Henderson, a technical officer with the Queensland Department of Primary Industries in Brisbane, contacted him for assistance.[6] She had the task of rearing crown-of-thorns larvae and was not having much success. She heard that Lucas had experience in larval biology and asked if he would have a try. She sent two jars of larvae to Townsville. The larvae would not develop for Lucas either. But when he raised the temperature, they developed and metamorphosed.

Lucas carried out this initial work secretly. He and others were advised by the head of the zoology department to have nothing to do with the crown-of-thorns because of its political sensitivity. But when funding became available through the crown-of-thorns Advisory Committee, he continued to study the growth and development of *Acanthaster* larvae. His results were incorporated in almost all models accounting for the infestations. In 1973, he showed that the larvae could not survive outside the temperature range of 24° to 32°C, and flourished in conditions of lowered salinity and higher temperature.[7]

Subsequently, Robert Pearson demonstrated that these sorts of conditions may occur within 50 kilometers off the North Australian coast where infestations were common.[8] The timing seemed just right. Spawning and larval development occurred during the monsoon season from December to February, when heavy runoff from rivers would lower salinity.[9] Pearson also examined the results of cyclone damage to reefs near Townsville, but found no evidence to support the hypothesis of Newman, Dana, and Fager that such damage was conducive to starfish aggregation. Nor did he find that most corals were killed.[10]

Other new evidence supported the view that the outbreaks were natural. The Advisory Committee pointed to the long-hoped-for geological studies of echinoderm debris in reef sediment that indicated the occurence of outbreaks before 1960.[11] In his preliminary studies, W. G. H. Maxwell had not been able to distinguish skeletal fragments of *Acanthaster* from the remains of other echinoderms. But in 1974, Edgar Frankel at the University of Sydney succeeded in establishing criteria for recognizing them.[12] He further made radiocarbon age determinations that, in the view of the Advisory Committee, "prove that significant aggregations of *A. planci* existed on various reefs at different times over the past 3000 years."[13]

The Advisory Committee concluded its report with a series of brief clear-cut statements:

> The crown-of-thorns starfish *Acanthaster planci* does not pose a threat to the Great Barrier Reef and in all probability it may never have done so.
>
> The observed increases in *A. planci* populations on some reefs during the past decade may not have been a unique phenomenon but rather one that had been more closely observed and more publicised than similar previous occurrences.
>
> The recovery of reefs destroyed by the starfish has been rapid and on most reefs no evidence has been found to suggest that any such reefs have suffered permanent damage.
>
> A mechanism which may cause population expansions in *A. planci* has been suggested. It involves the interaction of optimal physical, chemical, and biological factors whereby lowered salinity, adequate water temperature and good spawning occur at the same time, thus enabling a high rate of larval survival. These conditions occur from time to time in regions offshore from high land where heavy seasonal stream runoff leads to lowered surface salinities for short periods. This mechanism also accounts for the concentration of infestations on reefs near the land mass or near high islands and their comparatively rare occurrence on oceanic reefs.
>
> No evidence has emerged to support various assertions that man's in-

terference with the reef environment has caused increases in *A. planci* populations. On the contrary the evidence indicates that the *A. planci* phenomenon may be a natural part of the reef process and, in the long term, may contribute to the growth of the reef.[14]

CONTROLLING OUTCRIES

Members of the committee saw their job as complete, perhaps even unnecessary. They recommended that no more special funds be established for projects concerned specifically with the crown-of-thorns and any further research be financed through the Australian Research Grants Committee (ARGC).[15] Such research, they remarked, should "be determined on scientific merit and not in response to popular demand." [16] In making this statement, the committee pointed to a "special survey project" that, as they saw it, "originated more as an exercise in public relations within the political sphere rather than as a straightforward scientific survey." It concerned continued confusion over the actual extent of the infestations along with continued outcries for measures to stop their spread.

During the course of the Advisory Committee's operation, public attention on the crown-of-thorns did not subside. The previous Committee of Inquiry had rejected the survey of Endean and Pearson in 1966–67 on the grounds of incompleteness and selectivity in the sites surveyed. Yet the same group, as an Advisory Committee, rejected an application from Endean to conduct a new one. With help from other organizations and individuals, he and William Stablum undertook new surveys of eighty-two patch reefs between August 1969 and May 1971. They reported thirty-six reefs were infested and massive destruction on twenty other reefs that had been infested in 1966–67 but were now free of the starfish.[17] Warning that "additional reefs were coming under attack," and that the advancing front had moved southward from the central region of the Great Barrier Reef, they again called for measures to control the starfish plagues.[18] Their surveys, like earlier ones, were carried out with limited resources, used a variety of techniques, and were based on spot checks of small areas of reefs that had been reported to have large numbers.

Pressure was put on the Advisory Committee to support a more in-depth survey. In February 1973, it received a submission from the vice-chancellor of James Cook University stating that there was "an urgent need from the political, emotional, and scientific point of view for the present status of the crown-of-thorns infestations of the Great Barrier Reef to be assessed." Maxwell, who had left the University of Sydney to take up a position as technical manager at Australian Petroleum Exploration Association Ltd., was adamantly opposed.[19] But the committee was informed that "the Minister for Science wished to take some action that would allay recent publicity." Thus, with pressure from the government, new comprehensive surveys were conducted, with one of their aims "to demonstrate to the public that something was being done, and to counter much of the ill-informed criticism." [20]

Richard Kenchington was responsible for implementing the project. He was also partly responsible for initiating it. Like Lucas, Kenchington was strongly cautioned by colleagues on the hazards of conducting research on such a politically sensitive topic as the crown-of-thorns starfish.[21] Born in Barton on the Sea, Hampshire, England, Kenchington had completed an M.Sc. in marine biology at the University of Wales, specializing in plankton. He had moved to James Cook University in 1968. His interest in a new crown-of-thorns survey was not driven by attempts to allay public outcries. As a planktologist, he was seeking a large population whose larval cloud could be followed. In the early 1970s, he looked for the mega-populations alleged in the press to be on fifteen reefs off the coast of Townsville. But he could not find them. He muttered darkly about the need for an objective and comprehensive survey, and, as he recalls, his "bluff was called when the University's Vice Chancellor repeated his mutterings to a newly appointed Minister." [22] He soon found himself with three teams of Navy clearance divers, a landing craft, a patrol boat, an advisory committee, and instructions to design and carry out the survey.

Two surveys were conducted in 1973 and 1974. The Royal Australian Navy provided logistic support and additional funding. Members of the "Save the Reef Committee" directed by Endean participated in the surveys, along with Navy diving teams. Towed or free-swimming snorkel divers covered the entire perimeter of reefs in the region where *Acanthaster* was reportedly active. Based on comparisons with previous surveys, Kenchington reported that starfish concentrations had declined markedly. Only a few reefs at the extreme south end of the Great Barrier Reef seemed to carry major populations.[23] The results of the surveys were reported directly to the Minister for Science who released a press statement that "no widespread threat to the reef was posed by the crown-of-thorns starfish." [24]

In the view of some committee members, the Special Survey Project headed by Kenchington had simply been a waste of public funds. In Maxwell's view, "It added little to the understanding of the *A. planci*

phenomenon and one might well ask whether the effort, talent, and finance that went into this project could not have been employed more
effectively in fundamental research." [25] Actually, Maxwell went much
further and asserted that no special funds should ever have been given
for any *Acanthaster* research. Too many grant recipients simply did not
make significant contributions. And why was clear: there had been simply "insufficient people of the required calibre. . . . Had grantees been
obliged to compete with other branches of science through the normal
funding channels," he remarked, "the poorer-quality applicants would
have been eliminated." [26]

Even the most productive research, he argued, did little more than
confirm "the findings of the original Committee of Inquiry." [27] But if all
this were true, and research added so little to real knowledge, this then
begged the question: why was that committee so unsuccessful at closing
the controversy as it should have five years earlier? Large segments of the
public along with many biologists had judged that committee to have
been extremely heavy-handed in dealing with the issues. But, as
Maxwell saw it, the real problem was just the opposite: because the
committee wanted to appear unbiased, it was not critical enough of reports that the starfish was a human-induced ecological threat. In his
words, "In its concern for impartiality, the Committee of Inquiry may
have failed to emphasize its rejection of much of the dubious material
that had been given wide publicity during the early stages of the
A. planci controversy." [28]

By 1977, the waves of predators were disappearing, as adherents of
the natural cycle argument predicted they would. There was a collective
sigh of relief in the media, government, and marine science community
in Australia that the pestilence had proven a transient phenomenon.[29]
Scientists found it increasingly difficult to find specimens of the starfish
for research[30] and the controversy seemed over. But Maxwell did not
have the last word on what had caused it, what had gone wrong in the
relations between science and society to create so much turmoil. Any
complete assessment would have to include a more general evaluation
of the behavior of scientists in public forums, the role played by journalists, as well as the state of coral-reef science in the 1960s and 1970s.
This ecology demanded consideration lest history repeat itself and a
new uncontrolled controversy break out.

Richard Kenchington addressed these issues in 1978.[31] He began by
weighing the strengths and weaknesses on both sides of the controversy
over the previous ten years. He emphasized that the population
increases developed at a time when environmental matters were

becoming a major international political issue and were perceived as a threat to coral reefs "partly as a consequence of new technology, in the form of scuba diving, becoming generally available." [32] If the new technology partly caused the controversy, it could also help resolve it. Accordingly, he argued that it took some years "for field practices to be developed which employed the new technology effectively in sampling such a phenomenon over large areas." [33] Until then, proponents of the "man-induced" theory and ecological disaster had the upper hand.

The lopsided accounts of journalists selling fear and scandal heightened the controversy. Forecasts of disaster and claims of "whitewashing" and "vested interests" were "newsworthy," whereas the same could hardly be said for assertions of normality.[34] The case that the Great Barrier Reef was seriously threatened was also supported with photographs and films showing coral being destroyed. It was put fluently and emotionally, and backed by statistics.[35] On the other hand, the evidence against a serious "man-induced threat," Kenchington remarked, was "unconvincing in an emotional public debate" and "barely adequate in the scientific debates." [36] Thus, the controversy came to be represented "as a polarization of a 'pure' environmental lobby against official inaction and vested interests which were exploiting the Reef." [37]

The press did more than simply emphasize anthropogenic over natural causation. It presented the arguments in a simplified black-and-white way, selecting the more spectacular aspects of a complex situation and removing conditional clauses from scientists' statements. The potential threat to the reef was further enhanced by extrapolating from small areas surveyed to make major statements covering large areas of the Great Barrier Reef. For example, Pearson and Endean's original data on coral damage was transformed in *The Australian* as "more than 80% of the coral of the Great Barrier Reef." Further distortion occurred in the language deployed. The common use of the words "destroy" to describe the killing of living coral tissue incorrectly suggested that the starfish were eating away at the geological structure of the Great Barrier Reef.[39]

Acanthaster did make an ideal press story. But it would be false to attribute the exaggerations solely to ill-informed journalists searching for a story to peddle. Many of the "distortions" in media reports can be found in scientists' own reports in scientific journals.[40] That coral-reef erosion might occur was widely considered by scientists from the very beginning. In 1969, Jon Weber announced that, "the outcome could mean destruction for the entire Great Barrier Reef as well as many

Pacific islands between the Tropics."[41] And we should not forget Bernard Dixon's statement in the *New Scientist* that "more than a quarter of the total reef has disappeared."[42]

Nonetheless, Kenchington argued that scientists were perpetually at a disadvantage to nonscientists in the public debate "because they recognize that information is never complete in reality and that processes are usually complex." Coral-reef scientists, who in the 1960s regarded themselves as pioneer researchers working with incredibly complex systems they hardly comprehended, were suddenly asked by journalists and politicians for definite answers. Most suffered loss of face. In doing so, they appeared to the public to be uncertain and confused. On the other hand, "alarmists with a more direct and unqualified approach" seemed more convincing to the public.[43] Journalists had to rely on such "expert" testimonies. It was not easy for them to make extensive investigations themselves because access was difficult and costly, and any substantial investigation required scuba or snorkel diving.

But it was not just "alarmists" who behaved "unscientifically" and inflamed the controversy. So did members of the Committee of Inquiry. Kenchington debunked Maxwell's statement that "in its concern for impartiality" the Committee of Inquiry "may have failed to emphasize its rejection of much dubious material." It wasn't that they tried to be neutral. They were in no position to reject any claims. Endean and Pearson's report may have had some flaws, but the Committee of Inquiry had no data of their own to counter it, and their criticisms were hardly effective.[44] Without data of their own, Endean's position and his firm commitment to it, especially in the public debate, were more authoritative.[45] Moreover, as Kenchington saw it, he should never have been forced out. What was needed was a group of scientists, including Endean, to collaborate, in an attempt to devise a more comprehensive sampling procedure. Yet this was not done until 1973.[46]

Kenchington also confronted Maxwell's remarks that special funds should never have been allocated for *Acanthaster* research—that the monies should have been spent in other areas of scientific research. This attitude was indefensible. It only evaded the critical issue of directing attention toward areas of public interest that were scientifically undeveloped.[47] After all, the uncertainty over the crown-of-thorns reflected the lack of knowledge about normal ecological processes on reefs. Two interrelated factors accounted for this: the scarcity of marine biologists, especially coral-reef ecologists, and the high cost of marine research, particularly in remote areas.[48] The special problems and expense of research on remote reefs would have precluded almost all research into the crown-of-thorns without such *ad hoc* support provided by the special funds.[49]

Biologists in the late 1970s still complained that the most basic information on the population dynamics of the animal was lacking: detailed field studies of recruitment, lifespan, mortality rates, or major mortality factors of either adults or larvae.[50] This type of research required intensive and prolonged periods of carefully planned observation and experimentation in the field. Most dense populations of *Acanthaster* occurred on reefs that were typically 50 to 80 km from the mainland. It was necessary to spend prolonged periods on vessels, involving considerable constraints of expense, logistics, weather, and availability. With these difficulties, many interesting projects were beyond the reach of part-time researchers such as university faculty members, who had teaching commitments.[51] The real problem with allocating funds through the Crown-of-thorns Advisory Committee, Kenchington argued, was not a shortage of scientists of "required calibre," but a matter of research planning and management. More might have been achieved, he remarked, if specially funded research had been organized around a central, carefully managed, and well-serviced core program.[52]

Despite such problems, it was demonstrated that *Acanthaster* populations had declined, and Kenchington thought that the recovery of affected reefs would occur within twenty years.[53] He pointed to Frankel's studies of skeletal fragments indicating similar densities of *Acanthaster* populations dating back as far as 3000 years. In addition, Lucas's experiments showed that optimal larval survival occurred at temperatures that were higher and when salinity was lower than usual. "If a number of other factors such as food and predation are taken into account," Kenchington asserted, "it can be postulated that mass survivals of larval *A. planci* to the stage of settlement are natural but infrequent episodes in the long-term ecological balance of the Great Barrier Reef."[54]

The GBRMPA

Kenchington thought it unlikely that the occurrence of such a phenomenon affecting the Great Barrier Reef would ever again develop into so complex a controversy. In the first place, the crown-of-thorns controversy had led to a greater general understanding of reef processes. However, there was another major consequence. New policy decisions and investments had established a body responsible for reef conservation—the Great Barrier Reef Marine Park Authority (GBRMPA, pronounced "gebrumpa"), which possessed well-equipped and well-staffed laboratories as well as research vessels for reef research.[55] In

1978, Kenchington was appointed its Director of Research and Planning.

Legislation passed in 1975 made the Great Barrier Reef a marine park. That legislation was considered a great triumph for Australian environmentalists who had sketched a dramatic scenario of narrow interests of big business irrevocably destroying the great ecosystem. Oil drilling, mining, and other ecologically harmful activities were banned in perpetuity.[56] It is the largest marine park in the world; larger than Great Britain: 2000 km long, covering an area of 348,700 km^2 and comprising 2900 individual reefs, 300 reef islands or cays, 600 high islands, home to 1500 species of fish, over 4000 species of mollusks, and 450 species of corals. The unique environment of the Great Barrier Reef led to its inscription in October 1981 on the UNESCO World Heritage List.[57]

The GBRMPA regulated all recreational, business, construction, and research activities of the entire territory of the Great Barrier Reef.[58] Its management plan provided a model for juggling exploitation and protection of reefs in many countries. Over the next decade, tourism to the Great Barrier Reef attracted some 800,000 visitors annually. Many traveled to major new resort complexes. The most luxurious had an indoor ice-skating rink; the largest accommodated 11,000 guests. In the late 1980s, the area also featured the world's first offshore hotel—a seven-story structure built on a massive pontoon. The enormous increase in the human population along with the demands of shell collectors, divers, and commercial fishermen were regulated by a pioneering zoning system.[59]

Two "general-use zones" covered 97% of the park. The larger of these zones permitted commercial trawling and shipping; the other banned both and required permits for such activities as shell collecting and low-level aviation. Another 2% of the park was divided into three other zones that spelled out in detail what users, especially fishers, can or cannot do. One was a "look, but don't catch" area, another allowed only trolling for pelagic species, a third confined line fishermen to one handheld rod with a single hook or lure. The park's remaining 1% was cordoned off for scientific research, replenishment of marine life, and the preservation of areas in a state undisturbed by humans.[60]

In the 1980s, coral-reef communities around the world were estimated to cover about one million square kilometers (600,000 square miles). Most exist in poor countries where silt from erosion on logged-over watersheds often suffocated the coral. In other areas, sewage disposal, pollution, oil drilling, overfishing, dynamiting and other activities of ever-increasing human populations exacted high tolls. The rise of anthropogenic stress on coral reefs in many regions caused great concern for the scientists, governments, and public agencies. A major task of reef scientists was to make a correct evaluation of the situation on reefs and establish principles for a more beneficial relationship between humans and reefs. The GBRMPA also promoted educational work among tourists and local people, edited various booklets, and organized lectures and films. Many countries followed the Australian example and organized reserves or parks in reef zones. Their main goal was to offer the recreational and fish resources of reefs without doing harm to their ecosystems.[61]

Coral-reef research in Australia was also strengthened by the establishment of the Australian Institute of Marine Science (AIMS), in Townsville. It was created by an act of Parliament in 1972 and wholly supported by the Australian government. In 1978, AIMS had a staff of sixty-five researchers working in a main building with eight research modules, leading to a central library, aquaria, a lecture theater, and a boat harbor designed to accommodate two oceangoing research vessels and numerous smaller support craft. Thus, with the combination of the GBRMPA, AIMS, and James Cook University, Townsville became the major center for coral-reef research, teaching, and management.

Kenchington felt confident that such public controversy over the crown-of-thorns would be unlikely in the future. Ignorance, emotion, and vested interests would be overcome by the cold light of reason and the GBRMPA would resolve management problems. "In the event of the occurrence of an apparently anomalous or threatening situation," he wrote, "the Authority should be able to mobilize a large research team of considerable experience to investigate within a short time."[62] The notion that such public controversies may be quickly snuffed out by certain science may be the ideal. No matter how common that presumption may be among scientists and policy-makers, reality seldom, if ever, concurs. The crown-of-thorns controversy would prove no exception.

NEW-WAVE MANAGEMENT

At the time Kenchington published his paper, the momentum of public interest in *Acanthaster* died away, as did funds and the incentive for research. However, late in 1979, quite suddenly and inexplicably, the crown-of-thorns began to reappear. A new wave hit and it hit hard. In August 1979, Queensland Fisheries Staff reported large populations at Green Island. Estimates put the numbers as high as two million. It initiated a second series of infestations down the central region of the Great Barrier Reef and elsewhere over the next decade.[63] What was often heralded as "one of the strangest biological phenomena of this century" struck again throughout the entire Indo-Pacific region.[64]

The GBRMPA acted quickly and coherently, as Kenchington had promised it would. In 1979 it established an advisory committee; this time composed of scientists who had actually been involved in *Acanthaster* research, including Endean. The committee met in April and May 1980 and again in April 1984. Their duties were to review the results of research to date, assess whether new investigations were warranted and, if so, advise on possible management-oriented research.[65] The GBRMPA wanted advice on all the old problems: whether the outbreaks were a "normal" phenomenon or human induced, and what could and should be done to control them.[66]

Fieldwork had lagged behind laboratory work. The committee called for more research on the population dynamics of *Acanthaster* and the impact and behavior of large populations. Studies of fish populations, coral diversity, and community structure in "before" and "after" situations were called for. Such fieldwork might provide a basis for developing mathematical models of reef dynamics, which might contribute to the evaluation of available control measures.

Committee members were not optimistic about the effectiveness of existing control measures. Collection by hand was considered too labor-intensive and inefficient, even for small areas. The Queensland Fisheries Service tested toxic chemicals for injecting starfish as well as air inflation.[67] Of all the killing agents, copper sulphate seemed the most effective. The GBRMPA sponsored a trial control program around tourist viewing areas on Green Island reef in 1980. Two divers working for thirty-five days injected copper sulphate into the animals. They killed 25,850 starfish at a rate of 115 per hour. Nonetheless, "considerable numbers" still reached the tourist viewing area.[68]

The advisory committee requested further studies to determine whether there had been plagues prior to the 1960s. At its recommendation, members of the History Department of James Cook University conducted an oral history of the Great Barrier Reef.[69] They focused on recollections of old residents. There seemed to be a marked difference in the awareness of people of "European-cultural orientation" and those of "native-cultural orientation." In the European community there was virtually no awareness of crown-of-thorns between 1942 and 1960.[70] The indigenous community had a higher level of awareness, mainly because of the animal's capacity to "sting." They called it "Ur-me-meg"

("Ur" means "fire" and "mair-meg" means "carrier").[71] However, neither community associated the starfish with damage to the reef.[72] The
historians concluded that there was "no evidence that was compatible
with there having been a plague of crown-of-thorns at any time between 1920 and 1960, comparable in extent and intensity with that
which began in the early 1960s."[73]

Their results were in conflict with Edgar Frankel's sediment studies of
Acanthaster skeletons that had been taken as proof of past outbreaks.
However, his experimental results were not as decisive as some might
have thought. Endean criticized them in 1977, arguing that they only
demonstrated the presence of skeletal debris in sediments, not past
plagues.[74] First, there was no information on how many skeletal remains of *Acanthaster* per kilogram of sediment represent a past plague.
Second, the sediment tests seemed to conflict with field observations of
Acanthaster behavior. On Frankel's analysis, one would expect mass
mortality of *Acanthaster* following a "plague," thus resulting in a
marked increase in skeletal remains in sediments. However, no such
mass mortality had ever been observed after infestations.

The advisory committee also discussed possible anthropogenic
causes. The dramatic changes brought by commercial developments
and increased human populations in tropical areas had to be considered
in any assessment of reef ecology. That human activity may well have
upset a delicate balance on coral reefs that kept the starfish populations
in check was difficult to disregard. But one also had to consider ecological theory in regard to the nature of complex tropical systems.

10

COMPLEXITY AND STABILITY

The balance of nature has been a background assumption in natural history since antiquity.

F. N. Egerton, 1973

If a balance of nature exists, it has proved exceedingly difficult to demonstrate.

Joseph Connell and Wayne Sousa, 1983

Why are there so many species on coral reefs? What normally supports high species diversity? Are populations of species on coral reefs maintained in relative balance?[1] Are periodic large disturbances part of the long-term ecological processes on coral reefs? These issues entangled the heart of the scientific debates over whether the crown-of-thorns infestations were natural or human-induced, what the effects of the infestations were, and what should be done about them. Two competing conceptions underlay the controversy.

Some biologists assumed that complex reef communities were basically stable, tightly integrated systems in which a balance among specialized species was maintained by complex interactions that gradually evolved over eons.[2] According to this view, population explosions should not occur naturally, and it would take a long time for coral reefs to recover after major human-induced disturbances. *Acanthaster* infestations would involve many long-term, deleterious changes in the reef ecosystem.

Others denied that coral reefs possessed a particularly stable species composition. They interpreted ecological properties of reef communities as the result of largely random environmental disturbances, rather than as evolutionary equilibria derived from long periods of environmental stability. According to this conception, outbreaks of the crown-of-thorns would be one of similar kinds of natural disturbances. One could account for the apparent synchronization of the infestations throughout the Indo-Pacific as an artifact created by the great upsurge in recreational and scientific diving since the early 1960s. Damage caused by *Acanthaster* might even be part of the normal processes of reef development; and the ecological effects of the disturbances would be relatively short-lived (a few decades rather than many decades or centuries).

These conflicting conceptions partly reflected contemporary controversy in ecology concerning "the balance of nature." Ecology in the 1970s was in the midst of conceptual upheaval over the concept of "ecological stability" and its relations with complexity. Belief in the balance of nature goes back to antiquity. In its explicitly theological eighteenth-century form, for example, the harmony and order underlying nature's economy had a divine source. God's providence ensured a system of perpetual balance among all living things, in which each creature had its allotted place.[3] The balance of nature survived the emergence of Darwinian evolutionary theory in the nineteenth century. It was usually understood in terms of an integrated ecological system characterized by relations of interdependence and mutual benefit and regulating checks and balances driven by competition among individuals and species. As ecological communities became more diverse and integrated, they would become more stable.

SUPERORGANISMS

That species diversity leads to ecological stability had been postulated by many biologists for a hundred years. It was inherent in notions of evolutionary progress and was often applied to the human social world as well. It was central to the famous writings of the nineteenth-century British evolutionary and social theorist, Herbert Spencer and his followers. In Spencer's "synthetic philosophy," evolutionary progress in both the human social world and the natural world proceeded from the "homogeneous" to the "heterogeneous," from the simple to the complex. As long as competition and "that natural relation between merit and benefit" were maintained, there would be an increasing integration of all members of human society or the ecological community into a differentiated, mutually dependent, and efficient higher social organism. Violent competition and war would be replaced by a "peaceful competition" of "the free market." [4]

Spencer understood such mutual dependence and ecological integration in terms of the concept of the division of labor. Historians of biology have shown how the principle of division of labor was borrowed from the politico-economic theory of Adam Smith and went through various transformations under the pens of biologists.[5] The French zoologist Henri Milne-Edwards developed the concept of a physiological division of labor to describe the way the diverse organs of an organism carry out various functions for the benefit of the whole. Darwin subsequently used it to explain his theory for the divergence of species in terms of the advantages of evolving toward specialized ecological niches—a division of ecological labour. As he wrote in *The Origin of Species*, in 1859: "The advantages of diversification of structure in the inhabitants of the same region is, in fact, the same as that of the physiological division of labour in the organs of the same body." [6] Spencer took the analogy literally: the ecological division of labor, based on niche diversification, represented the differentiation of an ecologically stable superorganism. He elaborated his views in his treatise *The Principles of Biology*, in 1899.[7] "In the general transformation which constitutes Evolution," he wrote, "differentiation and integration advance hand in hand; so that along with the production of unlike parts there progresses the union of these unlike parts into a whole." [8]

From a top-down perspective of a general whole created by parts, Spencer could see mutual benefit and integration everywhere. Even predator–prey relations were not solely one-sided. One only had to view different levels of ecological organization to see benefit and look "beyond immediate results" to see "certain remote results that are advantageous." They brought about changes which "though injurious to the individual are beneficial to the species, and that, when not beneficial to the species, are beneficial to the aggregate of species." [9] Predators prevented "the inferior individuals—the least agile, swift, strong, or sagacious—from leaving posterity and lowering the average quality of their kind." They saved individuals who were feeble by injury and old age "from suffering prolonged pains."

Predators also put a check on undue multiplication. All was for the common good. Weasels not only benefit the plants eaten by rabbits and all those other animals that live on plants, but also rabbits themselves, since if rabbits were to increase beyond their means of subsistence, a large proportion of them would die of hunger. Such examples, Spencer argued, illustrated the numerous bonds that tied species together by mutual dependence so that one could "recognize something like a growing life of the entire aggregate of organisms in addition to the lives of individual organisms—an exchange of services among parts enhancing the life of the whole." [10] In the late nineteenth century, the concept of the division of labor was applied to all levels of biological organization. All "higher organization" was supposed to have evolved through the principle of the division of labor, reaching its fullest expression in the interdependence and mutuality of the constituent parts.[11]

Early leaders in plant ecology, Arthur Tansley in England and Frederick Clements in the United States, developed Spencer's superorganism concept. They maintained that competition among plants resulted in a highly developed division of labor in some plant communities, thereby producing a more integrated, differentiated, and stable adult state.[12] As Clements put it in 1916, "The life-history of a formation is a complex but definite process, comparable in its chief features with the life-history of an individual plant." [13] The concept that complexity implied stability persisted among leading ecologists throughout the twentieth century.

In the middle of the twentieth century, well-known works were written that contained arguments about stability, what determined it, and why its study is important. In *The Ecology of Invasions by Animals and Plants* of 1957, Oxford ecologist Charles Elton laid out the evidence that "the balance of relatively simple communities of plants and animals is more easily upset than that of richer ones; that is, more subject to destructive oscillations in populations, especially of animals, and more vulnerable to invasions." [14] The issue was important, especially for the agricultural practice of creating monocultures and using pesticide sprays. For if complexity and stability were correlated, Elton warned, "there is something very dangerous about handling cultivated land as we handle it now, and [it will be] even more dangerous if we continue down the present road of simplification for efficiency. . . . Invasions and pest outbreaks most often occur on cultivated or planted land—that is, habitats and communities very much simplified by man." [15] But the question of stability was important to the future of every species of the world as well as for methods of managing the world's ecosystems—not like playing chess; more like steering the boat, he thought.

Elton saw the argument that complexity leads to stability as needing research. However, he did point to indirect evidence based on research of simple communities. Predator–prey experiments on small animals in standardized laboratory environments showed that it was difficult to keep a small mixture of one predator and one prey in balance. Natural habitats in small islands, where species diversity would be relatively low, seemed much more vulnerable to invading species than those of the continents.[16] But for complex tropical systems there was only anecdotal evidence. Insect outbreaks, a feature of simpler temperate forests, were not known in species-rich tropical rain forests. As Elton recalled, "It was first brought home to me some years ago, when I had spent an hour expounding ideas about insect outbreaks to three forest officers from abroad. Then one of the men remarked politely that this question did not really concern them, because they do not have insect outbreaks in their forests! I found that he came from British Guiana, another from British Honduras, and the third from tropical India." [17] He emphasized that other ecologists speculated that in complex tropical communities "there are always enough enemies and parasites available to turn on any species that starts being unusually numerous, and by a complex system of checks and buffers, keep them down." [18] Elton commented, "Of course this is only a theory, and I expect only part of the story. But the ecological stability of tropical rain forests seems to be a fact." [19]

That increased ecological complexity gives rise to increased stability took a variety of forms. Some ecologists, such as Robert MacArthur, understood stability in terms of increased food-web complexity—that is, the diversity of pathways that energy could take through the food web in a particular community. As he put it, "Stability increases as the number of links increases." [20] In 1958, when Evelyn Hutchinson discussed the question "Why are there so many kinds of animals?," he raised the following metaphysical issues about the emergent properties of complex ecological systems:

> The evolution of biological communities, though each species appears to fend for itself alone, produces integrated aggregates which increase in stability. There is nothing mysterious about this; it follows from mathematical theory and appears to be confirmed to some extent empirically. It is, however, a phenomenon which also finds analogies in other fields in which a more complex type of behavior, that we intuitively regard as higher, emerges as the result of the interaction of less complex types of behavior, that we call lower. The emergence of love as an antidote to aggression . . . or the development of cooperation from various forms of more or less inevitable group behavior are examples of this from the more complex types of biological systems.[21]

If any system was characterized by such mutual interdependence, coral reefs with their great diversity of taxa and their numerous symbiotic associations were good candidates. The complexity and stability of coral reefs provided Endean and Chesher and those who followed them with a norm against which human interference could be assessed. Endean made the point explicit in 1977 when he argued that many ecologists regarded coral-reef systems as having a particularly stable or predictable organization because they are "biologically accommodated."

> Their trophic complexity and possession of a multiplicity of homeostatic mechanisms protect them against perturbations. The hypothesis that *A. planci* infestations are periodic or cyclic is not in accord with such views about the stability of coral-reef systems. For this reason and also because no valid mechanism for the periodic infestations has been proposed and no irrefutable evidence of previous population explosions of *A. planci* has been provided, the hypothesis that the *A. planci* infestations are periodic events appears untenable." [22]

Again, he insisted that "the virtual absence of scientific or historical records providing irrefutable evidence of previous *A. planci* infestations of reefs similar to those now occurring suggests that the *A. planci* infestations are unique. The catastrophic damage caused by *A. planci* infestations to reefs supports this view and this view is in accord with the widely accepted concept of the inherent stability of the biotic organization of coral reefs to normal (but not abnormal) perturbations." [23]

Endean was right. Such views about the inherent stability of coral reefs were widely accepted, especially by the Hutchinson school of ecology and the Caribbean school of coral-reef biology.[24] The view of complex ecological communities as species harmoniously adjusted to one another in a state of dynamic equilibrium was also perpetuated in popular magazines and nature films and commonly referred to by biologists. But it was very difficult to demonstrate. As Robert Paine emphasized in 1969, there was little or no sound evidence available to accept or reject such statements because "an operational definition of stability is lacking, as are data from the more complex associations." [25] The 1960s witnessed healthy debates over the concept of stability and the methods for verifying it—led by Paul Ehrlich at Stanford and Charles Birch at the University of Sydney.[26] Theoretical debates about what kinds of things coral reefs are persisted throughout the 1970s and 1980s. Why do tropical regions, by and large, support more diverse fauna than do regions of higher latitude? Were coral reefs biologically accommodated, co-evolved, highly organized and integrated species assemblages in which populations are regulated and species composition is structured? Were they "big organisms"?[27]

INTERMEDIATE DISTURBANCE

During the 1970s, questions about complexity and stability attracted the attention of many ecologists and evolutionists who sought more precise meanings for "stability." [28] Mathematical ecologists equipped with computers built models to test the relationship between complexity and stability. Their results indicated the opposite of what had been assumed: The more components in the models and the greater their complexity, the less likely the models were to be stable. Simple systems were usually more stable than complex ones. Beginning in 1971, the Australian physicist-turned-ecologist Robert May, at Princeton, began a systematic study of the question of stability in multi-species communities. His first paper began with the existing dogma: "One of the central themes of population ecology is that increased trophic web complexity leads to increased community stability." But May's conclusion was uncompromising: "We consider a simple mathematical model for a many-predator–many-prey system, and show it to be in general less stable, and never more stable, than the analogous one-predator– one-prey system. This result would seem to caution against any simple belief that increasing population stability is a mathematical consequence of increasing multi-species complexity." [29]

May expanded this theme in his influential 1973 book *Stability and Complexity in Model Ecosystems*.[30] He pointed to the crown-of-thorns to bolster the claim that complexity did not necessarily lead to stability: "Even the complex and diverse coral reef, commonly thought of as the aquatic analogue of the rain forest, has recently had its stability called into question by *Acanthaster planci* in the Pacific." [31] But, as he recognized, empirical evidence did not "yet permit a decisive answer as to whether complexity promotes population stability in the real world." [32] Many field ecologists of the 1970s defended the proposition that diversity promotes stability as one of the core principles of ecology.[33] Yet others went much further than May and questioned whether stability in the sense of an equilibrium among species ever existed.[34]

The notion that complexity, competition, and niche diversification resulted in the stable species composition and community structure of coral reefs was seriously challenged by Joseph Connell at the University of California, Santa Barbara. In 1962, he mapped and photographed several one-square-meter quadrats at Heron Island on the Great Barrier Reef; he returned year after year to see what had happened. His results were striking and had a profound effect on ecology. At least at the local quadrat level, the communities were unpredictable. The number of coral species and their abundance fluctuated greatly.[35] Connell suspected that the species composition of such communities seldom if ever reach a state of equilibrium. To the contrary, he proposed that high species diversity was maintained only when the species composition was continually changing. Again, he provided empirical evidence.

During the course of his studies on the Heron Island reef two damaging hurricanes had passed close to it, one in 1967 and another in 1972. Some areas were disturbed, others were not. When recolonization occurred on disturbed areas, there was an increase in the number of coral species. Connell noted that others had witnessed the same phenomenon: "Disturbances caused both by the physical environment and by predation remove coral and then recolonization by many species follows." [36] On the other hand, in those quadrats observed over several years that had not been disturbed, he noted a competitive elimination of neighboring colonies. For example, huge old colonies of a few species of staghorn corals overshadowed their neighbors.

Based on these and similar kinds of observations in regard to tropical forest trees, Connell proposed in 1978 that species diversity at a single location is best explained by a historical balance between exclusion of species by the best competitor on the one hand, and by disturbances that prevent this exclusion, such as storms, predators, or disease, on the other. His model, dubbed "the intermediate disturbance hypothesis," [37] predicted that diversity is higher when disturbances are intermediate on the scale of frequency and intensity. Diversity would be low at low levels of disturbance because the best competitor would become abundant and exclude other species. It would be low at high levels of disturbance because mortality would be too frequent, too recent, or too severe for many species to survive. Diversity would be highest at intermediate levels of disturbance—sufficient to prevent competitive exclusion by the best competitor but too moderate to eliminate most species.

Connell contrasted his model to the generally accepted view that coral reefs and tropical rain forests were stable ordered communities in which natural selection fits and adjusts species, and they return to their original state after perturbation.[38] Certainly, he recognized that coral-reef communities were more than a haphazard collection of species inhabiting a region that happened to tolerate the environmental conditions of the moment. Niche diversification by species played some role in maintaining local diversity. However, niche diversification unto itself, in his view, would not account for the high species diversity of coral reefs and rain forests.[39] The intermediate disturbance hypothesis became a key concept in community ecology. But it was far from clear whether Connell's results would hold for large spatial and temporal scales.[40]

Connell was certainly not the first to propose that disturbances such as typhoons and predation may act to enhance species diversity. After all, in the early 1970s, Branham, Porter, Newman, Dana, and Fager, among others, had suggested that *Acanthaster* predation may benefit coral species diversity. But, in 1978, Connell raised the crown-of-thorns as an example of just the opposite: a disturbance by predation that did not necessarily work to enhance coral diversity. Its predation did not prevent a few coral species from monopolizing space on a reef. Studies in Hawaii had shown that *Acanthaster* preferred relatively faster-growing corals (*Montipora*). However, they were not the most abundant species; they represented only 5% of the coral population around Molokai. Therefore, Connell argued that studies in Hawaii and those of Glynn in Panamá "indicate that the starfish attack rarer species preferentially, which would reduce diversity."[41] The effects of *Acanthaster* predation on other coral communities such as those in Micronesia and Australia remained unclear.[42]

The Effects of Keystone Predators

When Robert May emigrated from theoretical physics to ecology he noticed how ecologists engaged in healthy criticism of each others' work. "As a newcomer to ecology," he wrote, "I have been struck by the attitude of constructive interest in others' work which seems to prevail among ecologists. The competition and predation which characterize many other disciplines seem relatively absent, possibly because the field has not yet reached (or exceeded) its natural carrying capacity."[43] Whatever one might think of May's views on "the ecology of ecology" globally, they certainly did not apply locally in debates over *Acanthaster*. And whatever importance one might attribute to niche diversification, coherence, order, and stability in coral reefs themselves, these qualities were certainly not present in the research on *Acanthaster*.

In 1976, Roger Bradbury still complained of the lack of any directed accumulation of knowledge since 1969. The direction of research had been "almost random," and the most basic information on the population ecology of the animal was still lacking. One of the reasons for this, he suggested, was the "strange lack of scientific critique" and a "certain coyness which permitted anecdotal accounts to stand side-by-side with well conceived studies."[44] "The poor quality and undirected nature of much of the research," he wrote, "show clearly that we have refused to learn the scientific lessons, even though we were confronted by them.

The lack of critique shows that we have not even confronted the social lessons." [45]

Instead of studying the *Acanthaster* phenomenon, whatever its cause, to test basic assumptions of coral-reef ecology, disputants had merely locked the debate into natural versus anthropogenic cause.[46] The political context of the debate over controls gave specific meaning to these categories, and held them apart as mutually exclusive. Two symmetrical blocks of reason structured the debate. They can be grouped in three statements:

A. Outbreaks are novel.

B. The present ones are human induced.

C. Control programs are necessary.

Inverse:

A. Outbreaks have occurred in the past.

B. The present ones are natural.

C. One need not and should not control them.

This order of things—novel, anthropogenic, necessary controls, on one side, versus common, natural, and laissez-faire on the other—was not developed out of any inherent logical necessity; it was not determined by nature. It was a political order fashioned from an historical social context. To break out of this deadlock, and to test the strengths of competing hypotheses about the cause of the infestations and see alternative possibilities, the anatomy of the debate had to be deconstructed. Competing theories needed to be assessed in terms of their intrinsic logic, testability, plausibility, and simplicity, that is, the number of assumptions underlying them. Donald Potts offered the first detailed assessment of this kind in a lengthy overview published in 1981.[47] Taking on the task meant that he had to confront some of the sociopolitical forces that led to the opposing rhetorical paradigms.

In late 1976, Potts was approached by the editors of a volume on the ecology of "pests." They wanted someone with a marine background who was not involved in crown-of-thorns research to give an objective analysis of the literature. Potts, like Lucas and Kenchington, recalls that some marine scientists were reluctant to get involved.[48] Some feared that if their conclusions contradicted Endean's theory, they would be exposed to personal and professional attacks by him in the news media. Endean was considered powerful and influential; he socialized with important people and was one of the few who spoke with the press. Potts had been an undergraduate student at the University of Queensland in the early 1960s and remembers how the press would visit Endean's office with cameras. Endean was entrepreneurial; he received research funds to study toxins from pharmaceutical companies such as Roche; he was able to get funds outside the system at a time when most Australian academics were not. He got involved in gold mining and ranching; he was a millionaire. He was also from the country; he was okker, or played the okker (unsophisticated countryboy, the equivalent of a "good old boy" in the southern United States). Endean was also a Reader at the university. As Potts put it, "They were gods then"; they did little teaching, no administration. The University of Queensland resembled a British university of the 1920s. Most lecturers did not have Ph.D.s. Many of them went from being undergraduates to master's students to lecturers in the same university. Potts's was the first generation of students at Queensland to go overseas for a Ph.D.

In 1965, he attended the University of California, Santa Barbara, to study under Connell's direction. After completing his doctoral research on the comparative demography of land snails in 1972, Potts spent a year at Bishop's University in Quebec, before returning to Australia to work for a year at Flinders University in Adelaide. There, he became involved in coral-reef research. He frequently visited the Heron Island reef to study how natural selection affected coral in different habitats.[49] At the end of 1973, Potts moved to the Australian National University in Canberra where he remained for five years as a research fellow in the Research School of Biological Sciences. He realized that Endean was influential, had a long memory, held grudges, and that people were afraid that he was in a position to hurt them if they spoke up against him. However, Potts felt immune from such hazards. His position as a research fellow guaranteed ample funding for five years. So he had no fear of anyone close to Endean exerting pressure on grants committees. Moreover, Canberra seemed very far from Brisbane. Before he finished his assessment of the *Acanthaster* literature, in 1979, he and his wife, Laurel Fox, a specialist in plant-herbivore interactions, accepted a joint position at the University of California at Santa Cruz.

Potts's essay entailed a thorough critique of all the postulated theories, the nature of the hardened positions over the previous decade, and their underlying presuppositions. Ecologically speaking, three interrelated problems persisted in regard to *Acanthaster* plagues: their distribution, their cause, and the recovery of reef communities after infestations. Potts recognized that because of the extent of ignorance about so much concerning *Acanthaster*, it was difficult to make ecological inferences of the damage to coral-reef communities, their cause, and their recovery without assumptions about the nature of coral-reef communities.[50] To rid the problem of "emotionally charged" terms such as "plagues" or "infestations," he advocated the use of the relatively neutral term "outbreak" to describe all situations where unusually large numbers of starfish caused extensive damage.[51]

As late as 1979, Potts still found it necessary to assert that such damaging outbreaks of thousands of starfish persisting for months or years constituted "a very real phenomenon."[52] The number of starfish required to cause serious destruction differed in different reef contexts. And it was often difficult to distinguish between normal and outbreak populations in the literature. Potts thought that some of the reports of a few hundred starfish from New Guinea, Fiji, and West Samoa were actually of normal populations. Still other reports were anecdotal, and even when quantitative data existed, they were usually given as numbers seen in unspecified areas.[53] All major outbreaks with coral mortality exceeding 90% were confined to the Ryukyus in Japan, parts of Micronesia, and the Great Barrier Reef.[54] The few studies on coral regeneration after *Acanthaster* damage indicated that rates of recovery varied greatly among different habitats. In favorable situations, Potts suspected "species richness" might be restored in 2 to 3 years; in other habitats, recovery might take many decades.[55]

There were few data to refute any hypothesis for the cause of the outbreaks, whether it appealed to chemical pollution or to the postwar influx of scientists and tourists equipped with scuba. However, there were some general observations against which their relative strengths could be assessed. Any theory for the cause of the outbreaks had to consider that all major outbreaks were close to high landmasses, which were also centers of human activity, and their occurrence seemed to be synchronized throughout the Indo-Pacific. One also had to consider evidence for the existence of previous outbreaks. On these grounds, Potts evaluated the three main hypotheses: (1) "the predator-removal hypothesis" of Endean—based on overcollection of the giant triton; (2) "the juvenile recruitment hypothesis"—that outbreaks were caused by severe storms and terrestrial runoff leading to a combination of unusually low salinity and abnormally high temperatures; (3) "the adult aggregation hypothesis" of Dana, Newman, and Fager—that proposed no real increases in *Acanthaster* populations, but only redistribution and herding following severe storms.

"The predator-removal hypothesis," in Pott's view, was "the least satisfactory explanation on both theoretical and empirical grounds."[56] He argued that there simply was no demonstrated correlation between starfish and triton numbers on any reef. Furthermore, it was the most complex hypothesis, invoking numerous ecological assumptions: that the triton was a specialized predator of *Acanthaster*, that the triton itself was a "keystone" species that determined the structure of the reef community, and that reef communities are "biologically accommodated." Certainly, there were other known predators of the crown-of-thorns. Peter Glynn confirmed that the painted shrimp had the ability to reduce *Acanthaster* numbers in Panamá.[57] In 1973, Rupert Ormond and his collaborators reported the same to be true for the large puffer fish, *Arothron hispidus*, and the triggerfishes, *Balistoides viridescens* and *Pseudobalistes flavimarginatus* in the Red Sea.[58] However, the abundance of these predators and their effects on *Acanthaster* populations had yet to be examined in outbreak areas.

The triton control hypothesis as formulated by Endean was also the most rigid of the three hypotheses. Because it relied on the untested assumption of the inherent stability of coral reefs, and that they would resist *natural* perturbations, it rejected even the possibility of previous outbreaks.[59] There was little question that the evidence for past outbreaks was weak. Statements from residents, anecdotal references in the older scientific literature, and the folklore of some islanders conflicted. And Potts found it impossible to determine whether any referred to outbreaks. The only direct evidence for previous outbreaks came from Frankel's studies of skeletal fragments in sediments on the Great Barrier Reef. But there too the facts hardly spoke for themselves. Endean had rejected the conclusions because it wasn't clear how many spines in sediments represented a past plague and because field researchers had not noted mass starfish mortality during infestations that would presumably leave clusters of skeletal fragments in the sediment. Potts went further and denied the main presupposition underlying sediment studies in the first place: that evidence of the past outbreaks would provide evidence for present cause. "Recent outbreaks could have been caused by human activity," he reasoned, "even if earlier ones were initiated by other factors."[60]

Evidence of past outbreaks was as weak as evidence for the cause of present ones. But at least the other two main hypotheses for the outbreaks allowed for the possibility of previous outbreaks. They also supported the possibility that human activity was involved in contemporary outbreaks. Indeed, Potts made the strategic move of assessing them without regard to the intentions of those who had proposed them. Although they were used by their originators to deny that the outbreaks were caused by human activity and suggest that controls were unnecessary, these conclusions were not determined by the hypotheses themselves.

The "juvenile recruitment hypothesis," based on runoff, decreased salinity, and increased temperature leading to greater larval survival, invoked basically natural processes. But the same effects could be caused by human activities. As Potts saw it, the juvenile recruitment hypothesis actually allowed for "the probability" that outbreaks "may have increased by recent activity on land." [61] It predicted correctly that primary outbreaks would be associated with high landmasses that were more likely to receive heavy rains followed by massive runoffs than low-lying atolls. High islands were also more likely to support large human populations. This hypothesis also provided plausible explanations for the timing and distribution of the three major centers of postwar outbreaks as well as smaller outbreaks in other parts of the central Pacific. The initial outbreaks on the Great Barrier Reef and in Micronesia seemed to have followed a period of unusually severe typhoons that also coincided with rapid agricultural and residential development on land. Potts also noted that Japanese scientists believed that Okinawan outbreaks were associated with increased runoff from terrestrial developments.[62]

There were several problems with the juvenile recruitment hypothesis, however. At that time, no outbreaks composed mainly of juveniles had ever been detected. The hypothesis also had a number of special demands and required a tight sequence of biological and physical events. First, adult breeding aggregations had to be present at the right time and place so as to exploit the unusually favorable physical conditions. Yet, the hypothesis did not explain where the original aggregation of breeding adults came from. Second, spawning had to occur within a few days of a storm, if larvae were to develop while salinities remained low. Third, many larvae had to settle at the very site of the observed outbreak and not be dispersed away from the reefs. This was not a major problem in the confined coastal waters of the Great Barrier Reef. But around the isolated islands of Micronesia and the central Pacific, ocean eddies would be required to retain larvae for several weeks in the lee of the island before they settle. Finally, juveniles had to survive and remain aggregated for up to two years until they were large enough to cause sufficient damage for the outbreak to be noticed.[63] This was indeed a demanding scenario.

The "adult aggregation hypothesis," based on the feeding behavior of the starfish following coral-damaging typhoons, "provided a simpler explanation, especially for isolated oceanic islands." Dana, Newman, and Fager attributed the original coral mortality and resulting feeding ag-

gregations to storms. But Potts argued that that inference was not a necessary corollary of the hypothesis itself. The original coral destruction
could be the result of such anthropogenic disturbances as sedimentation, the result of erosion from marine and terrestrial construction and
agricultural sites.[64] The adult aggregation hypothesis also predicted the
possibility of previous outbreaks, and could explain the general distribution, local sites, timing, and age distribution of postwar outbreaks.
But it also had some obvious weaknesses. For example, weather records
at outbreak sites had not been compared with those from nearby sites
with normal populations, and such records probably did not exist for
much of the Indo-Pacific. But the most glaring weakness was inherent
in the hypothesis itself—that the outbreaks were composed of preexist-
ing adults. Almost all researchers in the tropics doubted that normal
dispersed populations contained sufficient numbers of individuals to
form the outbreaks of tens of thousands of individuals.[65]

To offset the deficiencies in the juvenile recruitment hypothesis and
the adult aggregation hypothesis, Potts suggested a hybrid. The adult
aggregation model would explain where the original breeding adults
came from, while the juvenile recruitment hypothesis could explain actual increase in *Acanthaster* numbers.[66] Similarly, the conflict over natural versus anthropogenic cause could be resolved and the two joined so
as to act together. Human activities may have increased the frequency of
the outbreaks. Although outbreaks appeared "to be responses to natural
disturbances," Potts argued, "indirect effects of agricultural, industrial,
or residential development on land may have increased probabilities of
occurrence of the hydrological conditions" for outbreaks.[67] In his view,
the dominant partner in the relationship was "nature, not man."

Finally, Potts addressed the heated issue of whether instituting control measures was necessary. Those who had argued that such
Acanthaster predation was normal and of benefit to the coral community had created their own functionalist teleology: if it's "natural," it is
desirable and "good." However, Potts recognized that *Acanthaster* predation was not responsible for maintaining coral species diversity: "the
hypothesis that normal *A. planci* populations maintain coral diversity
by preferentially feeding on common, competitively dominant species
can be rejected." [68] As an "essentially natural phenomenon," Potts
thought that the ecological effects may only be temporary. But if the
frequency of repeated outbreaks on particular reefs was increasing due
to human activity on land, he argued, "There may be long-term consequences." Potts thought that local control programs were justified at
tourist resorts.[69]

In the context of the politico-ecological debate in Australia, Potts may have wanted to temper the speculations of catastrophists. However, he underscored the point that the long-term consequences of the outbreaks could not be properly evaluated, "since ecological effects of *A. planci* populations (normal or outbreak) on the organization of reef communities remain almost unknown." All conclusions about the ecology of *Acanthaster*, especially those concerning the causes and effects of outbreaks, had to be regarded as tentative.[70] The debate over the cause and effects of the outbreaks remained wide open. New hypotheses were not long forthcoming.

New Hybrid

One of the major new theories for the cause of the outbreaks also relied on runoff from typhoons that bring heavy rains. But the main benefit to the larvae was not lower salinity and higher temperatures, but rather nutrients. It was proposed in 1980 by Charles Birkeland at the University of Guam, and independently corroborated by John Lucas. Beginning in 1984, Birkeland served on the Great Barrier Reef Marine Park Authority's crown-of-thorns advisory committee with Lucas. He and Lucas also wrote the first scientific book on *Acanthaster* biology: *Acanthaster planci: Major Management Problem of Coral Reefs*.[71] Initially, however, Birkeland was reluctant to get involved in research on *Acanthaster* outbreaks, but for reasons different from some others in Australia.

After completing his undergraduate degree at the University of Illinois, he moved to the University of Washington in Seattle to study under the direction of Robert Paine, who had invented the term and developed the concept of "keystone species."[72] Paine's focus on the special characteristics of individual species and their effects on community structure was to have an important influence on Birkeland's approach to *Acanthaster*. Birkeland's Ph.D. work, completed in 1970, was on a sea pen (*Ptilosarcus gurneyi*) and its seven species of predators.[73] However, during the next five years he received a remarkable introduction to coral reefs as a postdoctoral fellow at the Smithsonian Tropical Research Institute working under the direction of Peter Glynn.

There had been a large oil spill off the Atlantic coast of Panamá, and Ira Rubinoff sued the company and hired Birkeland and two others to examine the oil damage to coral reefs. It was understood that as long as they spent half their time working on the oil spill they could spend their remaining time working on problems of their own choosing.[74] As soon as Birkeland arrived, he had an exciting plunge into the growing coral-reef research community at a meeting on Glover's Reef off Belize, where a group of scientists met to discuss and write a major research proposal to the National Science Foundation. They included Ray Fosberg, Peter Glynn, Robert Johannes, Donald Kinsey, Judy Lang, Ian Macintyre, James Porter, C. Lavett Smith, Stephen Smith, David Stoddart, Frank Talbot, and Sir Maurice Yonge. They camped in grass huts on stilts, went diving every day, and wrote the proposal and talked into the night. Their idea was to request millions of dollars to set up tropical marine laboratories throughout the world to contribute to one huge computer model of coral reefs. Retrospectively unrealistic, Birkeland called it a "first-class boondoggle." But it was also one of the memorable events in the development of many leading coral-reef scientists.[75]

Birkeland's five years in Panamá were a spectacular experience—comparing the biology of the Caribbean coral reef with the nutrient-rich upwelling on the Pacific side 55 miles apart. The effects of nutrients, a major focus of his work in Panamá, became a major aspect of his research on *Acanthaster* outbreaks when he moved to the University of Guam in 1975. At that time, he considered the crown-of-thorns and the debates over the stability of coral reefs a "bandwagon." As a young scientist early in his career, he wanted to find his own topic rather than tag along on an exciting subject.[76] But his youthful individualism was soon washed away as he was swept up by a new wave of outbreaks.

In 1977, *Acanthaster* suddenly appeared with intensive outbreaks in the American Samoa and Palau, and a new outbreak occurred on Guam in 1979. In American Samoa, $75,000 was spent in an attempt to eradicate the starfish. After killing just short of a half a million around a tiny island half the size of Guam, it became apparent that they were not getting anywhere. Fisheries officials wrote to the University of Guam seeking scientific advice. Birkeland and Richard Randall responded. When they arrived in Samoa in 1979 Birkeland decided to be very methodical. He had no idea what the cause might be; he was going to look at fishing, the weather, predators, and check off things that just did not match.[77] Indeed, perhaps one of the strengths of his approach was that it was so naive, and so distant from the heated debates elsewhere.

The National Oceanic and Atmospheric Administration (NOAA) had tiny weather stations all over the island. They looked like "little out-houses" with a small dome on top; they were also located on Palau and Saipan. They had convenient single-page fold-up sheets for every year, with weather records going back to the turn of the century. As Birkeland recalls, the data, as it pertained to *Acanthaster*, popped right out. He saw a year of heavy rainfall, following a major drought. When he returned to Guam, he checked the records there. Again, there was a marked increase in rainfall in years close to the initial outbreaks. In both cases, the heavy rainfall occurred three years before the outbreaks. But that was exactly what one would expect, since as Birkeland reasoned, it would take *Acanthaster* three years to grow large enough to be observed. Guam and Samoa fit, but on different years. He then went to Palau and the data fit there as well; a drought had been followed by a severe rain storm and an outbreak three years later.

Like others, Birkeland noticed that primary outbreaks were almost exclusive to high islands, and that the spawning period of *Acanthaster* was at the beginning of the rainy season (June to August north of the equator, November to January south of the equator). This placed their larvae in the water at the very time they are most likely to be affected by heavy rains. The salinity and temperature conditions of the sea were important, but in Birkeland's view, they were not decisive. Nutrients would be a more predictive "key" factor.[78] When heavy rains follow dry spells, they cause terrestrial runoff of silt with nitrates and phosphates into coastal waters. These nutrients fertilize coastal waters and allow the occurrence of phytoplankton (algal) blooms upon which *Acanthaster* larvae would feast. Birkeland proposed, therefore, that "on rare occasions, terrestrial runoff from heavy rains (following the dry season or record drought) may provide enough nutrients to stimulate phytoplankton blooms of sufficient size to produce enough food for the larvae of *A. planci*. The increased survival of larvae results in an outbreak of adults three years later."[79]

The correlation of heavy rains following droughts with outbreaks of adults three years later was so good that Birkeland could state with confidence in 1982 that "all major outbreaks of *Acanthaster planci* have been associated with unusually heavy rainfall three years previous to their first appearance."[80] To support his view that the outbreaks recur naturally at irregular intervals he drew on Frankel's evidence of skeletal remains in sediment cores. He also noted that outbreaks occurred in areas far from agricultural activity or industrial and urban development (such as American Samoa and Palau). He collected further evidence of previous outbreaks on high islands from the cultural linguistics of people in Micronesia, Polynesia, and Melanesia. People from high islands had traditional cures for punctures from *Acanthaster* and species-specific names for the starfish. In Palau, *Acanthaster* was called *rrusech* while other starfish are called *btuch* or *tengetang*. At Fiji, it was called *bula* (a homonym of "hello") while the general terms for starfish are *gasagasan* or *basage*. In the Cook and Society islands, *Acanthaster* was called *taramae* and in Samoa and Tonga it was called *alamea*. Yet, people from atolls did not remember previous outbreaks and referred to *Acanthaster* with names that were general names for starfish.[81] On their first trip to American Samoa, Birkeland and Randall interviewed local fisherman who informed them that there had been two other outbreaks there. One was near the turn of the century and one in the 1930s. They said the starfish were so common that they would not fish at night for fear of stepping on them. Older fishermen on other High islands, the Solomons, New Ireland, Pohnpei, and Palau, told similar stories.[82]

At James Cook University, Lucas had already turned to investigate nutrient requirements for larval development. In the early 1970s, when he studied the effects of temperature and salinity on *Acanthaster* larva survival and development, he did so with seawater enriched with cultured algae. He considered his work on temperature and salinity as an initial step because these were the simplest parameters to manipulate. However, since *Acanthaster* breed during the summer when water temperature is favorable, temperature itself could not explain sporadic increases in larval survival. The larvae tolerated relatively low salinities, but he thought salinity was also unlikely to have a big enough effect on survival levels to be a major factor. This led him to consider the other potentially important factor that could be manipulated in the laboratory: food availability.[83]

Certainly other factors affected larval survival, such as predation and unfavorable dispersal, but these required field studies that were extremely difficult to carry out. Coral reefs usually exist in oceanic conditions that are low in nutrients and phytoplankton. So fluctuations in phytoplankton, the starfish larvae's food, seemed a potential mechanism for major variations in larval survival. By 1980 Lucas had acquired the experience and equipment for doing quantitative studies on larval nutrition. He raised larvae on several species of unicellular algae and found that they developed more slowly or starved to death when provided algae in concentrations normally found in Great Barrier Reef waters.[84] Only in particularly dense blooms were algal phytoplankton abundant enough in nature to provide adequate food for mass larval survival. His results fit nicely with the model Birkeland coincidentally developed.

Birkeland became a vocal critic of the argument that coral reefs were naturally stable systems with populations approximately in equilibrium.[85] Tropical insect populations were assumed to fluctuate less than temperate insect populations on the basis of the diversity–stability hypothesis, but Birkeland could point to new data indicating that plagues of tropical insects were related to rainfall pattern and added nutrient for larvae.[86] Like others, he also emphasized that many marine inverte-brates with planktotrophic larvae were characterized by great year-to-year fluctuations in populations. Outbreaks of *Acanthaster* had merely received more attention than had outbreaks of other species be-cause of their spectacular effects.[87]

By 1983, Birkeland could list other large-scale tropical outbreaks that had not captured environmentalists' concern to the extent that *Acanthaster* did.[88] That year disease caused the mass mortality of 99% of the *Diadema antillarum* urchins on the Caribbean coasts of Panamá and Colombia. The previous year, there was a widespread killoff by dis-ease of *Echinothrix* urchins from Kauai to the big island of Hawaii. Dis-eases causing mortality of several species of corals had affected reefs in Bermuda, Florida, St. Croix, and Central America. Epidemic diseases had killed several species of sponges in many areas of the Caribbean and the Gulf of Mexico at least six times in the hundred years following 1844. In 1983, seventeen million birds comprising eighteen species had disappeared from Christmas Island. Were all these events repercussions of large-scale human activities, an historically recent factor in the tropi-cal marine environment, or natural phenomena? Birkeland suggested that the disappearance of sea birds from Christmas Island may be the re-sult of shifts in major oceanic current patterns called El Niño. El Niño had occurred eight times in the previous forty years and the sea bird populations had apparently recovered each time.

One issue was certain: one could not assess the cause of such changes on the premise of naturally stable communities in the tropics. To ex-plain that error, Birkeland pointed to the sporadic way scientists had studied tropical environments. Until the 1960s, most marine laborato-ries were in temperate regions, and tropical marine communities were investigated by biologists attached to expeditions.[89] The population dy-namics of conspicuous long-lived marine animals in the tropics (corals, sponges, giant clams) occurred on time scales that were beyond the range of studies during expeditions. On the other hand, animals with short life spans were generally cryptic or under cover, and their popula-tion dynamics were not obvious.[90] Thus, Birkeland predicted that the belief that large-scale fluctuations of tropical marine populations were unnatural might change with long-term studies by residents.[91]

Birkeland suspected that much of ecologists' intuition of tropical ecology had been built on temperature. Annually, temperatures vary less in tropical regions than in temperate regions, so it was naturally assumed that populations also fluctuated less in tropical environments. Other environmental factors such as availability of chemical nutrients in the water column and vagaries of oceanic current patterns were difficult to measure. Again, he was hopeful: "Now that marine laboratories are established in the tropics and data are being obtained by resident scientists an inductive picture of tropical marine populations undergoing dramatic fluctuations as a result of a degree of reproductive success which is significantly influenced by changes in nutrient availability may replace the deductive model of a biologically accommodated steady-state community." [92]

Birkeland's theory for *Acanthaster* outbreaks was fundamentally based on natural causation. However, like Potts, he emphasized how deforestation in tropical coastal areas could increase the frequency of their occurrence. As he wrote in 1983,

> Although the nutrient runoffs that caused the outbreaks of *Acanthaster* to date have not clearly been a result of human activity, we should take heed that if land on hillsides in tropical coastal areas is cleared for forestry or agriculture or urban development at an increasing rate in the future, the increased erosion of nutrients into the ocean because of clearing vegetation from the topsoil could result in increased frequencies of *Acanthaster* outbreaks. [93]

During the 1980s, Birkeland emphasized more and more that the outbreaks may well have been exacerbated in frequency and magnitude by such coastal developments. [94]

Public attitude toward the crown-of-thorns in the United States settled to a negative, but not alarmist position during the 1980s. There was not much pressure on government or academic institutions to establish programs for eradicating *Acanthaster*, but sport divers around Guam would perfunctorily ream the center out of any they came across. Diving clubs also organized eradication programs when herds encroached on their favorite dive sites. [95] In the Ryukyus, diving instructors at hotels or those leading tours formed regular removal programs. Larger starfish populations were sometimes handled by larger cooperative efforts. In 1983, near Sesoko Island, the township of Motobu, the University of the Ryukyus, the staff of a commercial aquarium, and local fishermen all cooperated in a removal program. [96] In Australia, where such control programs were lacking, public outcries again rose like a phoenix out of the ashes as a new wave of outbreaks made their way down the Great Barrier Reef.

11

CYCLICAL OUTCRIES

It is unlikely that the occurrence, in the future, of a phenomenon such as *A. planci* affecting the Great Barrier Reef, would develop into so complex a controversy.

Richard Kenchington, 1978

The outbreaks on the Great Barrier Reef reached the newspapers in Australia in the early 1980s with new allegations of a conspiracy by the "authorities" to conceal the extent of starfish damage and their unwillingness to do anything about it.[1] Such articles as "Reef Killer Plague," and "Return of the Coral Killers," informed readers that the second major plague, possibly worse than the first, was "being covered up by a combination of silence and ignorance."[2] The Melbourne newspaper, *The Age*, raised the question: "Are we heading for The Great Barren Reef?"[3] There were renewed claims that tourist operators were trying to keep the public in the dark about the new "plague"—of "monumental proportions . . . so big it makes your mind boggle."[4] Endean reportedly stated that "nobody is saying anything because they don't want to stir the possum—I would say this present infestation is much worse than the first one." At the same time other marine ecologists told reporters they were reluctant to become involved because of the political aspects.[5]

The return of the outbreaks in 1979 seemed to vindicate those who had warned that they were not simply a passing nuisance that would go away on their own. Theo Brown, director of the Australian Division of the World Life Research Institute during the 1980s, told readers of *The Australian* how he had been "subjected to ridicule and abuse from opposing academics" when he "accurately documented the serious threat posed to the reef." [6] He explained how in 1972, following a survey by the Save the Reef Committee, over seventy divers were available to institute starfish controls and thus to reduce the threat. But, their proposal was "*effectively* vetoed by senior academics from the administration of James Cook University and no subsequent action was possible."

Brown explained scientists' construction of natural causes theories as due to "anomaly anxiety": "The conventional researcher trained in an area of expertise tends to experience this when faced with new evidence that doesn't fit the prescribed parameters of his or her knowledge." [7] The plague should not have occurred; it was contrary to all past documented knowledge and teachings in marine science, Brown wrote. Therefore, the phenomenon either had to be denied or explained away. He himself was characterized as "a liar by unnamed academics" when he reported outbreaks in 1969. "However, when faced with the overwhelming evidence of coral-reef destruction, and perhaps out of sheer desperation, the opposing academic group spawned the natural cycle hypothesis to explain away the phenomenon." [8] Brown pointed to marine pollution, the overcollection of the giant triton, and overexploitation of some species of reef fish. Although more research was needed, he argued, there was no longer any valid argument to delay the introduction of starfish controls.

The GBRMPA (Great Barrier Reef Marine Park Authority) was at the center of the new assaults by those demanding action. When addressing the outbreak of one to two million starfish at Green Island, the GBRMPA's Chairperson, Graeme Kelleher, wrote in *The Australian* that "A total of more than seven man-years of effort would be required to kill this number." Brown charged that Kelleher's statement was simply designed to mislead the public. "While it could conceivably take one man seven years to eliminate one to two million starfish," he asserted, "a team of 100 trained divers, working only an eight-hour day, would have the job completed in 10 to 25 working days. Perhaps it's time for those scientists who have repeatedly slated Dr. Robert Endean, Bill Stablum, Dr. Peter James and myself as alarmists to realize that Dr. Germaine Greer's famous eight-letter word can best describe their argument that the crown-of-thorns starfish should be left unchecked to ravage the corals of the Great Barrier Reef." [9]

If the plague of starfish was not larger than the previous one, the public outcries certainly seemed to be. But this time, the GBRMPA launched its own public campaign with new data and new experts. John Lucas responded in *The Australian* and explained that there was no compelling evidence for either human-induced cause or natural fluctuations. [10] The disagreement among marine biologists was due to the fact that they simply did not fully understand coral-reef ecosystems and the factors producing changes in animal populations on coral reefs. Calls for controls, he argued, were often made in ignorance of what was feasible. He pointed to major control programs of little benefit, costing tens or hundreds of thousands of dollars, in Japan, Micronesia, and American Samoa. Starfish remained after the money ran out and, in time, the large populations declined, whether they were attacked or not. Even then, those control programs were directed at starfish on accessible fringing reefs. The logistics of tackling large populations on the distant offshore patches of the Great Barrier Reef were much more daunting. On the other hand, Lucas recommended keeping tourist coral areas free of the starfish. He asked the public for patience in waiting for answers from research, especially field studies of population dynamics. He assured readers that statements made in the 1960s that it might take many decades for reefs to recover from the starfish predation "would not now be made in the light of the acquired knowledge of coral reefs."

In January 1984, the Minister for Home Affairs and Environment, Barry Cohen, replied to media reports of "massive reef destruction by the crown-of-thorns starfish, and that government departments were playing down the situation." [11] Such claims of massive reef destruction, he retorted, were oversimplified and exaggerated; research and surveys would continue. In 1982, the GBRMPA initiated a user-report system to collect and analyze survey and sighting reports of *Acanthaster*. It constructed a survey questionnaire and distributed it widely to scientists, dive clubs, dive shops, and cruising charters. It also reviewed data from surveys carried out since 1966. Records had been analyzed from 516 individual reefs out of a total of approximately 2500 in the Great Barrier Reef region—a sample of just over 20%. [12] Of those reefs examined over the previous eighteen years, starfish had been "common" or "found in clusters of more than forty starfish" only on 23%. Cohen commended the efforts of tourist operators to control starfish populations on the reefs they use. The GBRMPA would provide technical advice on the most effective ways of controlling starfish numbers in small areas. No control methods existed other than injecting individual starfish with a toxic chemical or collecting them by hand. They were both ineffective and labor intensive for large areas. Individual treatment techniques, he argued, could eliminate a maximum of about 140 starfish per dive hour; and because some reefs have been assessed to have in excess of two million starfish these methods were impracticable.[13] Emphasizing that the weight of scientific opinion was that the outbreaks were cyclical and natural, Cohen compared killing great numbers of starfish "to killing large numbers of kangaroos because they are eating grass."[14]

In a press release the next month, Graeme Kelleher explained how the GBRMPA had contributed $302,000 toward research relevant to the incidence and distribution of crown-of-thorns starfish on the Barrier Reef.[15] Insisting that the starfish plague was not as widespread as the press claimed, he pointed to the results of new surveys of 151 reefs showing that only about 14% of them had aggregations of forty or more starfish.[16] He also asserted that "recovery had been reported in many parts of the tropical Pacific over the past 20 years."[17] It would be "irresponsible to recommend the commitment of substantial public funds in order to meet demands that something should be done, while there is every reason to doubt the need for or the effectiveness of such action."[18]

Under the title "March of the Starfish,"one could read in bold print—"The Great Barrier Reef is being destroyed by crown of thorns starfish and the Federal Government does not care." And in small print, "Or the starfish are a natural phenomenon of the reef which are not causing any more damage than they ever have before."[19] Statements from the GBRMPA asserting "no increase in starfish numbers and that the Reef is re-growing as fast as it is destroyed" were pitted against those of Endean that the current plague, "the worst in history," had already destroyed 80% of the Great Barrier Reef, stretching down the Queensland coast, and it "could face total destruction within 10 years."[20] The GBRMPA was also under suspicion in the media of being tied to the commercial interests of tourist operators concerned that the starfish publicity would damage their industry.[21]

To counter such perceptions, the GBRMPA and the Far North Queensland Promotion Bureau invited media representatives in early February 1984 to visit different reefs off Cairns to show them firsthand the "true situation" and answer questions on the destruction of the coral and natural regeneration of affected reefs. The GBRMPA prepared an information kit designed to promote better understanding of the crown-of-thorns and its impact on the reef. Several members of its advisory committee went on television and radio to allay fears, while others met with media representatives.[22]

The GBRMPA's efforts to provide the media with a more realistic and hopeful perspective had an immediate, but not long-lasting effect. Under the headlines "Crown of Thorns—No Threat to Reef," "Crown of Thorns Problem Not as Bad as It Seems," "Starfish Claims Rejected," and "Science Gives Life Back to Reef,"[23] journalists reported that the GBRMPA was well on the way to finding the solution to the sudden starfish outbreaks. At the same time, others dismissed the trip to the reef by national news media representatives as simply an attempt by the Far North Queensland Promotion Bureau to "whitewash the problem." Tourist operators who wanted the government to pay for control programs to protect tourist reefs insisted that "if the Government sent small teams of divers down it would be easy to control starfish plagues on the tourist reefs. It was only a matter of spending money."[24]

Other scientists entered the public debate on radio programs to criticize the GBRMPA. Anne Cameron, Endean's former student and colleague at the University of Queensland, charged that Kelleher's claims about the extent of the destruction and recovery were misleading. She reproached him for extrapolating from a survey of some hundreds of reefs, of which only 14% carried the starfish in large numbers, to give a consoling picture about the overall picture of the 2500 reefs of the Great Barrier Reef. Indeed, scientists who had conducted surveys of forty-nine reefs in 1980 warned that it was impossible to make categorical statements about the size and movement of the populations on the Great Barrier Reef as a whole or even provide "a true picture" of what was happening in the section of the Great Barrier Reef they examined.[25] Cameron also found it hypocritical that the GBRMPA would supply some tourist operators with equipment and advice on how to control starfish on their little patch of reef, while asserting that it does not want to interfere with what Kelleher called a "natural process."[26]

A CONSPIRACY OF SILENCE?

In 1984, journalist and filmmaker Robert Raymond published "A Conspiracy of Silence over the Destruction of the Reef" in *The Bulletin*[27] (the Australian equivalent of *Newsweek*). He spent several months with divers and a film crew examining the impact of the starfish on reefs, talking to boat captains, dive leaders, tour operators, and resort proprietors. He was convinced that the infestations, this time, were far more severe than the ones he had seen in 1969. "The damage to many of the most accessible reefs is so serious," he wrote, "that, for those who have yet to see the Great Barrier Reef, I have only this advice—choose your area carefully. And do not wait too long." [28] Diving and snorkeling boat captains called one reef "Atomic bommie" because it had looked like an atomic bomb had been dropped.[29] The threat "that the authorities had tried to keep the lid on for 30 years," Raymond wrote, was now showing signs "of blowing up into the biggest environmental problem the federal government has faced." [30] The impact on the booming tourist industry promised to make Australians as familiar with the crown-of-thorns as they were with rabbits, locusts, mice, or any other plague-prone animal.[31]

Raymond questioned Richard Kenchington, Director of Research and Planning at the GBRMPA. Kenchington produced figures to support the GBRMPA's position that, of the reefs for which it has reports—about 14% were infested with starfish.[32] He dismissed the large-scale starfish control programs tried in other Pacific territories, such as Micronesia and Okinawa, as merely "PR management stunts" that achieved very little"; they were "a great palliative exercise and you feel that you're doing something." He further emphasized that although the GBRMPA had provided some people in Queensland with injection devices, it was "not actually in the business of picking up the tab for collecting starfish." Graeme Kelleher explained government policy that when there is something required purely for a local enterprise, then it is the enterprise's responsibility to meet the cost, just as, generally speaking, it was a farmer's responsibility to keep the dingoes or the rabbits down on his property.[33] Kelleher assured Raymond that there was no evidence that the Great Barrier Reef was at risk.

At James Cook University, Lucas told of his latest research on larval survival: why in some years only one or two of the twenty to thirty million eggs produced by each starfish survive, and why in others one thousand from each female go on to reach maturity. He explained how after a particularly wet rainy season washes nutrients into the sea, there would be more food and greater survival of the larvae. But, as Raymond pointed out, further testing of that hypothesis would have to be done by someone else. Lucas had completed his laboratory work on the requirements for *Acanthaster* larval development and had turned to study the biology of giant clams.[34] Field studies, especially population dynamics, continued to lag behind laboratory studies. Raymond interviewed the director of the Australian Institute of Marine Science (AIMS), John Bunt, who informed him that scientists were unable to work on the population dynamics of the starfish because of the technical difficulty of tagging the organisms. *Acanthaster* had a certain "plasticity" that enabled them to discard anything tied onto them. In 1984, Raymond reported that he could find no scientists in Australia working specifically on what caused the starfish outbreaks.[35]

He also spoke with Endean and "listened once again to his expression of anger and frustration at what he calls our lost opportunity to do something about a national environmental disaster." Endean clung to his belief that the outbreaks were due to overfishing of starfish predators, not just the giant triton but also large reef fish, coral trout, wrasse, various cod, and the grouper. He insisted that the new wave of outbreaks was far worse than the first. In the previous wave, mostly branching corals and plate corals were affected. But this time he found that the hemispherical coral, the huge brain corals that took hundreds and in some cases thousands of years to grow to their present size, were being killed. He prophesied that there would now be "a succession of infestations, each one worse than the preceding one." He pointed a finger at the previous crown-of-thorns committees for not launching eradication programs earlier when they had the chance:

> We have had the opportunity to halt these infestations and not taken the opportunity. We've let a very conservative scientific establishment dictate to our governments and say that there's nothing to worry about. It's obvious now that those people were wrong. It's not just a passing phase; it is something which is really significant and something which is causing massive damage which may well result in permanent impoverishment of the Great Barrier Reef.[36]

Adopting Endean's perspective, Raymond elaborated his views about the "cover-up" and lack of efforts to control the starfish plagues in his 1986 book, *Starfish Wars: Coral Death and the Crown-of-Thorns.*[37]

New scientific evidence supported the views of those who criticized the GBRMPA for making reassuring extrapolations and downplaying the extent of the new wave of outbreaks. Extensive surveys conducted in 1985 and 1986 suggested that the outbreaks were more widespread and more serious than the GBRMPA had publicized. They were carried out by the "Coral-reef Ecology Group" at AIMS, led by Roger Bradbury, Peter Moran and Russell Reichelt. In 1985, they received about $A1 million from the Commonwealth Community Employment Program, GBRMPA, and the Queensland National Parks and Wildlife Service to further their research. Their research program covered four main areas: population dynamics of the starfish, dynamics of coral community, dynamics of predator–prey interactions, and the ecosystem context.[38]

With special funds from the Commonwealth Community Program,

they hired a large number of students and unemployed biologists to help survey 228 reefs. Their results indicated that the outbreaks were about twice as widespread as the GBRMPA previously claimed: 28% of the reefs surveyed had recently experienced or were experiencing an outbreak of the starfish.[39] The results were especially cause for concern, particularly to managers, because the regions of reefs most affected were within the central third of the Great Barrier Reef, an area of extensive tourist-related activities. It was clear to Bradbury, Moran, and Reichelt that "the crown-of-thorns starfish has had a greater impact on the Great Barrier Reef in recent times than any other known man-made or natural disturbance." [40]

Publishing their new data and disseminating it was far from straightforward. As Bradbury recalled, the new director of AIMS, Joseph Baker, was horrified by these numbers. Baker, a natural-products chemist, was also a member of the GBRMPA and good friends with Kelleher, who had been the target of so many public attacks. However, the data from the new survey were difficult to refute. Not only were more reefs assessed than by any previous study, but this was a rigorously stratified random sampling. Bradbury, Moran, and Reichelt wrote their paper in 1985, thinking it was appropriate for *Nature*. AIMS was a government-run institution and research results were to be sanctioned by the director before being made public. Normally this was a formality. In this case, the director dragged it out over about eighteen months, constantly asking for minor editorial changes such as recommending the word "perturbation" be changed to "disturbance." It was a very tense time, especially for Bradbury.[41]

Bradbury had long been aware of the complexity of environmental issues since the late 1960s when, as a graduate student at the University of Queensland, he became involved in the "Save the Barrier Reef" campaign. His thesis supervisor was James Thomson, a fish taxonomist who had been a member of the joint government crown-of-thorns Committee of Inquiry. But Bradbury himself developed away from traditional natural history. He completed his Ph.D. in 1973 based on a study of fish-community structure in Morton Bay entitled "Complex Systems in Simple Environments." He had also acquired considerable academic experience before arriving at the Australian Institute of Marine Science. After a year at Dalhousie University, Nova Scotia, as a postdoctoral fellow with the well-known population ecologist E. C. Pielou, he spent two years at the University of Sydney working with Australia's leading terrestrial ecologist Charles Birch, a champion of non-equilibrium approaches to population biology, who believed that weather drove most insect populations. Bradbury spent another two years at the University of Wollongong as lecturer, and then a year and a half at the Office of National Assessments in Canberra (a government institution responsible for examining the strategic position of Australia, geopolitically, economically, and technologically). By the time Bradbury moved to AIMS in 1979 he had acquired a thorough knowledge of ecological theory, and would become well known among coral-reef ecologists as a "mathematical wizard." Although he began as a community ecologist, he turned to modeling with computers, as he put it, "the bigger the better." [42]

In those tense months of 1985 and 1986, Bradbury was sensing a "cover-up." Their paper was not published until 1988. In the meantime, the GBRMPA adjusted its public statements about the extent of the outbreaks. In 1985, its crown-of-thorns advisory committee reconvened and conceded, albeit clothed in caveats, that the starfish did pose "a serious threat to the organization and functional relationships within some reef communities within the Great Barrier Reef, at least in the short term." [43]

THE GBRMPA VINDICATED

By 1986, the Australian government had spent more than $3.5 million on *Acanthaster* research. [44] Nonetheless, Kenchington still found himself in the same situation as the crown-of-thorns committee in 1970. Managers who were expected to act to avert, eliminate, or contain possible catastrophes could not refute statements by experts such as Endean and Cameron that the situation was indeed catastrophic. [45] Based on studies in the early 1980s, Kenchington reported that it would take ten to fifteen years for a seriously affected area to reach levels of coral cover similar to those on unaffected reefs. [46] Reestablishing a similar range of species was more complex. [47] In the early 1980s, the GBRMPA commissioned a report on methods for accelerating the recolonization of corals. It found that the "coral garden" approach of transplanting entire colonies or large fragments to damaged areas was successful and appropriate for small areas. [48]

There were a few records of successful eradication programs. In Micronesia, intensive control seemed to result in an overall decrease in a starfish population after a two- or three-year period. [49] However, one also had to take into consideration the natural longevity of the populations. On the Great Barrier Reef, starfish populations decline markedly when the animals are four years old. [50] Bradbury and his colleagues at AIMS also argued in 1985 that control programs would not lead to large-scale eradication of the starfish and would have little, if any, effect on the occurrence of future outbreaks.[51] Others emphasized risks in instituting large-scale control programs. By thinning the population densities of *Acanthaster* and keeping the numbers of *Acanthaster* below the level at which they devastate their food resource, control programs might be effectively functioning as "prudent predation," producing an "optimal yield" of *Acanthaster*.[52] This issue was raised in Okinawa in 1986 by Masashi Yamaguchi, who suggested that the chronic situation in the Ryukyus may be partially due to the extensive control programs there that thinned populations but did not eliminate them. Japan had spent more than 600 million yen (about $US400,000) to remove over thirteen million *Acanthaster* from the Ryukyu Islands between 1970 and 1983.[53] Yet, in 1986, Yamaguchi considered it "safe to assume that most reefs in the Ryukyus have been devastated in the past 15 years." [54]

Virtually all scientists agreed with the GBRMPA that efforts should be concentrated on the protection of selected sites of particular importance for tourism, science, or unique resources.[55] Hand collection was shown to be successful in eradicating small starfish populations of 500 to 1000 individuals and thus saving small areas of coral. In such small-scale, specific-site, control programs, there would be no overall counterproductive effect on the population dynamics of the starfish. Without knowing the cause of the outbreaks there was little room for optimism regarding effective controls. As Kenchington put it, "Control of starfish populations and the enhancement of recovery of affected areas may be likened to treatment of symptoms of an unknown condition. Where they can be applied they may provide some immediate relief and gratifying action." [56] The key strategy to long-term management, he argued, lay in research on the cause and effect of the phenomenon.

In 1985, the GBRMPA's advisory committee recommended a five-year coordinated program of research on the management and ecology of the starfish, at an estimated cost of $A3 million. The program included testing theories of the cause of the outbreaks, more trial eradication programs on selected reefs, studies to determine when controls were likely to be feasible, socioeconomic studies of the effect of outbreaks on tourism and other reef industries, and public education on the problem and what the government was doing about it. The GBRMPA's attempt to manage the relations between science and society continued through video productions and publications aimed at laypeople.[57]

Public criticisms did not subside until 1988 when GBRMPA and AIMS were brought before the public with accusations on the Australian news-magazine television program *60 Minutes* that they were underestimating the extent of the outbreaks and not doing enough to remedy the situation. As a result of the television show, the Minister for the Arts, Sport, the Environment, Tourism and Territories requested a review to be conducted of crown-of-thorns research and management programs. After consultation and correspondence with Australian scientists and those from other countries, the direction of the research programs and the GBRMPA's management policies were endorsed.[58] It would not be long before a new complex theory for the cause of the outbeaks was introduced, one that took account of both natural reef processes and human activities.

12

CROSSROADS

Despite what some arm-chair theorists may have us believe, the problem is far from solved. Indeed it may take many more years before a clear understanding of the crown-of-thorns starfish phenomenon is achieved, if in fact this is possible. Unfortunately, it seems that there is still a significant number of individuals who prefer the comfort of simple answers to simple questions . . . At present there are no simple answers to the crown-of-thorns starfish phenomenon—only an array of questions about a rather complicated biological problem.

P. J. Moran and R. H. Bradbury, 1989

. . . the roots of the problem may go far beyond the ecology of *A. planci*. There is no cause for complacency.

C. Birkeland and J. S. Lucas, 1990

LONG-DISTANCE ECO-EYED VIEWS

The world was faced with a mass of urgent environmental problems by the late 1980s: global climatic change, acid rain, overfishing, destruction of tropical forests, effluent and pollution of all sorts, resistance of pests and pathogens, epidemics of new diseases. Ecologists were anxious because the public in general, and decision-makers in particular, still seemed to regard them "as fuzzy remnants of flower power, not really relevant today." [1] Yet one of the results of studying community ecology, ecologists argued, was that one could have prior understanding of the likely environmental risks in a particular situation.

Partly to persuade society that their skills might be of considerable value, a group of ecologists met in 1987 at the Asilomar Conference Center in California. Robert May, Paul Ehrlich, and Stuart Pimm discussed ecological stability and the difficult issue of scale: that things look very different depending on the point of view. A cow sauntering in a meadow sees a sward of roughly equal edible plants. A butterfly flitting across that same meadow looking for one plant on which to lay her eggs views things very differently. To her, two plants just centimeters apart are not equivalent. But ecologists were less certain than cows and butterflies about what scale they should observe nature to understand it best.

Time is a major factor. Watching some systems for a long time, whether the Pacific climate or a small field, sometimes answered ecological puzzles. In doing so, Ehrlich explained, ecologists were "beginning to understand more and more that very significant events can go on at a time scale of 50 to 100 years."[2] Things bumble along in their ill-understood fashion for many years and then suddenly something happens that makes sense of everything that went before. Ehrlich studied a species of checkerspot butterfly, *Euphydryas ethida*, in an area where it lays its eggs on only one plant, a species of Indian paintbrush, *Castilleja linariifolia*. There were other plants around, plants that the butterfly uses in other parts of its range, but in that one area Ehrlich studied, it stuck to the one plant for its larvae. Why? "It took us a very long time to realize that the reason that the butterflies used only that plant is that about every 50 or 100 years there is a severe drought in the area, and the species they used was drought-resistant and therefore would carry them through those years. So there was a very strong evolutionary pressure for the butterflies to remain on that plant, but you would never discover what that was unless you happened to have had, as we did, an opportunity to observe a drought year."[3]

For another example, Ehrlich pointed to the crown-of-thorns starfish. It also showed how a little ecological application could bring things into focus. He explained how "interventionists" believed that human activities had triggered the outbreaks in some way; they tried to isolate the cause and in the meantime went out onto the reef to round up as many starfish as possible. Others, whom he called "real scientists," examined cores drilled from the reef. "They found the remains of these starfish in sediment about every 200 or 300 years," Ehrlich explained. "So this was a regular thing that happened on a cyclic of two or three hundred years, but this was the first time it had happened since scuba gear, and so it was the first time anybody had noticed it."[4]

Robert May had argued that most often what we observe in nature will be stable systems. Stuart Pimm from the University of Tennessee spoke of diverse aspects of stability. "Persistence": when a system lasts unchanged for a long time relative to the life spans of its component species. "Resilience": a measure of how quickly the system returns to its initial state after a disturbance away from equilibrium. "Resistance": a measure of how well the system copes when a variable is permanently changed.[5] The plagues of starfish represented an example of one of the diverse forms that stability took. Coral reefs were stable in the second sense. They were thought to be resilient and would return to equilibrium in a relatively short period.

The ecologists at Asilomar Conference Center were not as confident about the cause of some other cases of dramatic population fluctuations. For example, the sardine fishery off the California coast suffered a total collapse after 1944. Ehrlich believed it was due to overfishing and hence the population had failed to recover. "The entire structure of the community changed so that the sardine could not come back."[6] May was less certain. He pointed to the herring fishery in the North Sea and the Baltic that collapsed dramatically in the 1980s. Although people were quick to blame overfishing, he pointed to "a certain amount of evidence over the past several centuries for cycles of collapse and recovery of herring fisheries."[7]

While the ecologists were not certain why that fishery collapsed, they were convinced that ignoring ecologists had already cost governments dearly. Ehrlich told the story of the Mediterranean fruit fly—the notorious "medfly." When it invaded California's orchards, politicians responded to farmers' pleas with a massive spraying program costing about $100 million. Yet, it may all have been a waste of money. There were three types of medfly: those from Costa Rica, those from Hawaii, and those from Germany. If the California invaders had come from Germany, then the orchards were in big trouble because that strain could survive the winter. But if the medflies had come from Hawaii or Costa Rica, a single California winter would have killed them. What appeared to have happened, Ehrlich argued, was that "we had a huge spray program, badly carried out, but the fly was actually wiped out by the cold weather that would have come anyway." It was not just the waste of money that concerned Ehrlich, but the possible danger to people from the widespread use of pesticides. "Ecologists knew how to go about it," he asserted, "but they were ignored."[8]

The uncertainty of ecology, ex-physicist May argued, was overplayed. There was

> nothing special or unusual about the character of ecological theory. Many people have a feeling that the physical sciences are characterised by a crisp determinacy, and that biological predictions having to do with resource management are simply ineluctably more fuzzy. There is some truth to this, but at the same time there are areas of physics . . . that are

every bit as ill-understood as many biological things. And conversely there are many things in biology that are quite well understood.[9]

Again, he pointed to the North Sea fisheries, where ecological understanding was sufficiently precise to give adequate practical advice. Implementing that advice was another matter. "Problems that are often mistaken for problems of uncertainty are really political and social problems of translating reliable advice into action."[10]

As Pimm has emphasized, most ecologists deal with processes that last for only a few years, involve no more than a handful of species, and cover an area of a few hectares.[11] Yet the most pressing ecological problems involve many species and their fate across decades to centuries. In 1981, Paul and Anne Ehrlich predicted a catastrophic loss of species over the next decade that would rival the mass extinction 65 million years ago at the Cretaceous–Tertiary Boundary when the world lost half its animal species, including dinosaurs and many lineages of plants.[12] Managers of national parks and conservation areas were charged with protecting many species and ten to thousands of square kilometers against a spectrum of possible threats. For them to manage, ecologists had to understand how populations and communities would change over the very temporal and spatial scales about which they knew the least. In 1991, Pimm compared the importance and complexity of their problems to "the stupefyingly tedious sequencing of the human genome":

> With complete certainty, I predict that there will be human genomes around in fifty years to sequence; with somewhat less certainty, I predict that there will be at least ten billion of them, dying from many causes each of which is orders of magnitude more important than the genetic causes the human genome sequencing will uncover. If we do not understand ecological processes better than at present, these ten billion humans will be destroying our planet more rapidly than we are now. When we contemplate this, it is no wonder we ecologists take ourselves off to beautiful, untouched environments and study fascinating species.[13]

The importance of large-scale and long-term ecology was recognized by many ecologists. Yet theoreticians building large-scale ecological models were often criticized for simplifying things and for their lack of data. No one could do neat controlled ecological experiments at the level of complexity and temporal and spatial scales involved. Therefore, modelers had to be opportunistic, scavenging data from various sources to test their models.[14] To put it bluntly, the criticism of this approach was that "the models are terrible and that the data are even worse."[15] Ehrlich's use of *Acanthaster* outbreaks to support his notion of

two-hundred-year cycles is illustrative. It also reveals considerable confusion among ecologists who viewed the crown-of-thorns controversy
from a distance.

ON THE WATERFRONT

Certainly, the favored explanations for the outbreaks relied on behavior
phenomena of the organism and its relations with the physical environment, salinity, temperature, and food. Nonetheless, the cause of the
outbreaks still remained a subject of controversy among those who actually engaged in research on the outbreak fronts in Australia and Micronesia. No coral-reef ecologist who studied the crown-of-thorns in
the late 1980s would have agreed that the observed plagues were the result of two-hundred-year cycles. Edgar Frankel's sediment data appeared to indicate a rough clustering of the radiocarbon ages of previous
aggregations at 250- to 300-year intervals. This was certainly consistent
with Richard Chesher's arguments about the old age of some of the
coral killed. However, Frankel suspected that this clustering may well be
an artifact of the techniques he used for dating *Acanthaster* spines.[16]
Birkeland and Lucas had taken Frankel's data simply as evidence for
past outbreaks.[17] The view that these outbreaks were due to nutrient
runoff, phytoplankton blooms, and increased starfish larval survival
following severe storms was incompatible with two-hundred-year cycles. The occurrence of two major outbreaks since the 1960s also conflicted with this notion.

The second major wave of outbreaks of the 1980s also lent considerable weight to the view that they were not merely natural fluctuations.
In the late 1980s Birkeland repeatedly emphasized that they may well
have been exacerbated in frequency and magnitude by human developments.[18] The outbreaks recalled by elder fishers occurred at intervals of
several decades—two or three times a century. But they appeared now
to be more frequent and were sometimes chronic. For example, no outbreaks were remembered on Guam before the 1950s, but there was a
major outbreak in 1968 and another in 1979. The second outbreak had
perpetuated itself for a decade and continued. Similarly, *Acanthaster*
had become a chronic problem in the Ryukyu Islands beginning in
1959; it too persisted.

The crown-of-thorns outbreaks were not the only tropical marine
phenomenon that had increased in frequency. Birkeland pointed to
poisonous "red tides" and paralytic shellfish poisoning caused by
dinoflagellate blooms that had been increasing at a geometric rate in the
western Pacific since 1975. Dinoflagellates (Greek *dinos*: whirling) are

unicellular algae that are among the most important photosynthetic organisms in marine ecosystems. Some species of this group swim freely as part of the biodiverse plankton grazed on by fish. Other forms managed to get inside corals, sea anemones, and clams, evolving over eons an intimate symbiotic relationship, supplying food made from photosynthesis in return for protection and nutrients in the otherwise nutrient-poor waters of the tropics.

Some of the dinoflagellate species that form red tides produce a toxin that attacks the nervous systems of fishes, causing massive fish kills, especially in the tropics, that are deadly to humans as well. Paralytic shellfish poisoning causes respiratory failure in humans who eat oysters, mussels, or clams that fed on certain dinoflagellates. Increase in the frequency of red tides, caused by population explosions of dinoflagellates, was likely caused by runoff of nutrient-rich human and animal wastes. The clearing of tropical forests for agricultural and urban use had accelerated in the previous twenty years, increasing the runoff of nutrients and sediment in coastal regions. It was not difficult to imagine that crown-of-thorns outbreaks, due to increased food (phytoplankton blooms) for starfish larvae, were rising in frequency because of coastal developments in southeast Asia and the western Pacific. Therefore, Birkeland argued, reducing the frequency and magnitude of the outbreaks would require careful management of developing coastal lands.[19] Birkeland's model was highly regarded by coral-reef ecologists, but there was still another aspect of human activity in coastal areas that could not be ignored: overfishing.

OVERFISHING

The effects of overfishing the starfish's predators had long been considered. In his 1981 review, Potts regarded the predator-removal hypothesis as the least satisfactory. However, during the second half of the 1980s, the coral-reef ecology group at the Australian Institute of Marine Science (AIMS), led by Roger Bradbury, Peter Moran, and Russel Reichelt, reevaluated it, and all other hypotheses, in light of new arguments and further evidence.

Between 1985 and 1989, the Australian federal government provided about $2.5 million for a coordinated research program on *Acanthaster* and coral-reef ecology.[20] It was a large, multidisciplinary, multi-institutional effort involving the collaboration of seventy scientists from throughout Australia and overseas in fifty-eight different projects. While the Great Barrier Reef Marine Park Authority coordinated projects that were largely management-related, biologists at AIMS coordinated ones that were mainly ecological. Their studies of the crown-of-thorns were to have a bearing on many other aspects of large-scale coral-reef ecology. As Moran and Bradbury wrote, "The uniquely large scale of the phenomenon provides scientists with the perfect natural experiment: a window through which they can observe, and test, many important concepts in coral-reef ecology which ordinarily they would be unable to investigate." [21]

In reassessing the main theories for the cause of the outbreaks, ecologists at AIMS intended to move the debate away from existing dichotomies—natural versus anthropogenic—and their false rhetorical conclusions in regard to cause and controls. Their aim was concentrated on the poorly understood predator–prey relations. They recognized that many coral-reef ecologists supported the view that the outbreaks were essentially natural and could be exacerbated by the increased runoff from human developments on coasts. But the acceptance of that view was merely that. All the evidence supporting it was embedded in interpretation.

Moran, Bradbury, and Reichelt argued that there was no direct scientific evidence to suggest that outbreaks occurred prior to the late 1950s—when they were first reported in the Ryukyu Islands between Japan and Taiwan.[22] The prominence of the crown-of-thorns in the memories and cultures of indigenous peoples might only be due to the fact that they are conspicuous and venomous.[23] Frankel's sediment data, they argued, merely provided evidence of the existence of *Acanthaster* over the past several thousand years, not of outbreaks.[24]

Despite all the logical criticisms of evidence based on the sediment record, there were new and improved sediment studies in the late 1980s designed to show that *Acanthaster* outbreaks were common in the past. They were conducted by Peter Walbran and Robert Henderson and their colleagues at James Cook University.[25] In their report in *Science*, they argued that large *Acanthaster* populations had been part of the Great Barrier Reef for at least 8000 years. They further attempted to revitalize the long-discussed and often rejected notion that *Acanthaster* outbreaks would enhance coral species diversity. Referring to Connell's "intermediate disturbance hypothesis," they concluded that such disturbances would result in an "enhanced capacity for speciation and the generation of unusually high diversity levels." [26]

It was not long before these new studies were criticized and rejected by biologists at AIMS. They did so on much the same grounds as previous sediment studies. But there was one important additional issue. The geological studies assumed that there would be a straightforward stratigraphy, one would be able to detect a disproportionate number of spines in different layers in the sediment record. In doing their tests they bunched spines from a certain layer together, assuming they were the same age, ground them up, and dated them. But this assumption was faulty. AIMS had on staff a biologist interested in taphonomy: the study of the conditions at the time of an organism's death that led to the preservation of its remains in the record. He pointed to a huge literature on sedimentology. In turn, biologists at AIMS highlighted the activities of *Callianassid* shrimp that burrow in the sediment in the lagoon around the base of reefs in search for food. As a result of their burrowing, and siphoning water in and recycling it, the shrimp constantly turn over the top two meters of sediment. Because of the shrimp's activity there would be no way of getting any sense of when spines accumulated. There would be no way to distinguish whether a clump of *Acanthaster* spines resulted from an outbreak or from their being sifted through the sand. Therefore, they argued, no correlation could be made between the abundance of skeletal elements and starfish population size past or contemporary.[27]

That the novelty of the outbreaks was merely due to the advent of scuba and the increased use of coral-reef environments for tourism and research was also pure conjecture. As Moran asserted, "It is no doubt true that these factors have been responsible for our greater awareness of the distribution and abundance of *A. planci*; it is however sheer speculation to suggest anything more than this." [28] In an extensive overview in 1986, he argued that all existing hypotheses for the outbreaks were based on correlations; none of them demonstrated true cause and effect, and all had flaws.

Anne Cameron and Robert Endean had raised doubts about the validity of Birkeland's hypothesis on the grounds that if *Acanthaster* outbreaks were caused by terrestrial runoff, why were there no outbreaks of some other starfish with similar life histories?[29] To resolve this issue, further studies of the larval ecology of other asteroids would have to be carried out. There were also uncertainties about the actual weather conditions and the time of initial outbreaks. Although the terrestrial runoff hypothesis might hold for isolated areas in Micronesia and the South Pacific, Moran argued, no adequate data existed on weather conditions and initial outbreaks in the Great Barrier Reef to determine whether the hypothesis could be applied there as well.[30]

Matching the timing of outbreaks to predicted weather conditions was not easy for any theory. For example, Birkeland reported that an intense cyclone crossed the Queensland coast in 1959; according to his theory, this would have followed a dry period.[31] Yet, as Dana, Fager, and Newman note, increased cyclone activity was experienced along much of that coast during the period from 1958 to 1961. Their own hypothesis had similar problems.[32] Matching cyclone activity on the Great Barrier Reef with herding behavior was not the most serious criticism of the adult aggregation hypothesis, however. Many ecologists who actually observed the outbreaks found it absurd to suggest that the hundreds of thousands, sometimes millions, of starfish seen on some reefs could ever have arisen by the aggregation of a normal population of dispersed individuals.[33]

Little was known about predator–prey relations, and Moran argued that overfishing could not be dismissed. Overcollection of the giant triton as a cause of the outbreaks had been criticized since the late 1960s.[34] In the early 1980s, Endean extended his hypothesis to include the effects of fish predators such as the grouper *Promicops lanceolatus*.[35] Of course, he had also embedded his hypothesis in the premise that outbreaks had not occurred in the past because coral reefs are biologically stable and predictable systems. Regardless of that assumption, it was still possible that the removal of fish predators played an important role in the outbreaks. Perhaps the removal of predators and the increased larval survival due to runoff acted together.

Moran offered the following scenario: Adults may aggregate under natural conditions. If the spawning of these adults coincided with times of heavy runoff, high food abundances, and optimal physical conditions, then that might lead to increased survival of larvae. However, the settlement of large numbers of larvae and the establishment of dense aggregations of juveniles would occur only if predation was not extensive.[36] The important point, he argued, was that "this phenomenon may not be explained easily and that to trust one hypothesis is akin to putting on blinkers." The possibility that *Acanthaster* outbreaks may be causing irreparable changes to many of the world's coral reefs also remained.[37]

As Bradbury recalls, during the 1980s, they were not thinking that the outbreaks were caused by overfishing. It seemed that they might be locked into some sort of predator–prey cycle, like that of hare and lynx. They would not need any external factor, they were "self-perpetuating, tightly linked, just lock and step and roll on."[38] AIMS biologists concentrated on consequences more than causes: studying the nature of predation, recovery, and testing Connell's intermediate disturbance hypothesis.

That was before 1988, prior to Rupert Ormond's visit to AIMS.

Ormond and his colleagues had been studying fish predation on *Acanthaster* in the Red Sea since the early 1970s. In 1973, when they reported that both the large triggerfish and the large pufferfish were predators of large adult *Acanthaster*, they suggested that if these fishes responded to initial increases in starfish populations by feeding on them more frequently, they might be capable of keeping populations in check.[39] But it was not known if they were important predators on the Great Barrier Reef. In subsequent feeding experiments in the Red Sea, they found other suspects that were part of the commercial fisheries. They showed that several species of Emperor Bream (family *Lethrinidae*) fed readily on small echinoids and asteroids.[40] They also observed them feeding on juvenile *Acanthaster*, and in 1988 research conducted through the GBRMPA found *Acanthaster* remains in the stomach of an Emperor Bream.[41]

Based on standard ecological population equations, Ormond and his collaborators developed a computer-based model to predict threshold densities of those predators needed to keep the starfish population in check. They tested their model by comparative surveys of the predators in the Red Sea where no major outbreaks occurred and those in the Great Barrier Reef. In the Red Sea, densities of the pufferfish, triggerfish, and lethrinids were found to be well above the predicted threshold. In July 1988, they surveyed ten reefs in the central and northern sectors of the Great Barrier Reef for the same starfish predators; they were below the predicted threshold and their numbers were lower on outbreak reefs than non-outbreak ones.[42] There were also fisheries data showing that the total fish catches in the area had increased tenfold during the early to mid-1960s to a plateau sustained during the 1970s and 1980s. Separate records were kept for one species of lethrinid, the redthroat, *L. chrysostomus*, a preferred target of reef anglers and spearfishermen. The records indicated that the catch of this species decreased rapidly during the 1960s.[43] All this supported their view that intensified fishing pressure since the 1960s may have facilitated the current series of *Acanthaster* outbreaks.

RIP

The next month, the sixth International Coral Reef Symposium was held in Townsville; most of the world's coral-reef scientists attended. The coral-reef ecology group at AIMS seized the opportunity to run a workshop on the crown-of-thorns immediately before the symposium. They invited a small band of mathematicians including Peter Antonelli from the University of Alberta and Nikolai Kazarinoff from the State

University of New York, Buffalo, who had been modeling the phenomenon,[44] to meet with scientists actively working on the phenomenon in the field. The workshop resulted in a volume edited by Bradbury, *Acanthaster and the Coral Reef: A Theoretical Perspective.*[45] In introducing it, he announced "an emerging synthesis that reflects a far deeper understanding of this complex phenomenon than was possible even a few years ago."[46]

Ormond, Bradbury, and their collaborators constructed a synthetic model of how the starfish may be regulated by fish predators.[47] However, they were careful to include a number of conditions. First, the presumed fish predators would feed opportunistically on a wide variety of invertebrate prey and only switch to feed heavily on any one prey type, such as *Acanthaster*, when that prey population increases. Second, all the water conditions caused by runoff for increasing larval recruitment were needed. Third, threshold numbers of predators would not always prevent outbreaks; exceptionally high levels of recruitment could overcome it. Thus, once outbreaks become endemic in a region, the extent of larval recruitment could be so high that it would swamp the regulatory effects of crown-of-thorns predators. Therefore, occasional outbreaks would be expected in an earlier historical period prior to significant impact by humans.[48]

Thus, overfishing could be only one of a series of factors that interact to determine *Acanthaster* population levels. To contrast their model "with oversimplistic 'larval recruitment' or 'predator removal' hypothesis," they proposed the term "Recruitment Initiated Predation" (RIP):

> The term is intended to emphasize that our model anticipates that significant predation of juvenile and sub-adult starfish will only occur following heavy local recruitment of COT larvae. The RIP acronym is also intended as a reminder of the fact that unless a more open-minded attitude to understanding the COT phenomenon can prevail, continuing degradation of the GBR and other reef areas may lead to their irreversible decline.[49]

Combining different conditions in diverse ways also meant that diverse paths to outbreaks were possible. Outbreaks would not be due merely to singular events, such as heavy runoff. They would result from dynamic processes with discontinuous effects. Bradbury teamed up with Antonelli and adopted the catastrophe theory as a framework to embrace various conditions.[50] Their model suggested three distinct outbreak paths. However, they emphasized that "the present evidence points to the recruitment initiated predation hypothesis (RIP) as the most likely explanation for the current series of outbreaks on the Great

Barrier Reef." [51] RIP was not immediately accepted by all coral-reef scientists, but, in Bradbury's view, it was accepted by enough of those who studied *Acanthaster* to be considered a new paradigm. [52]

CONVERGENCE

Acanthaster and the Coral Reef: A Theoretical Perspective was rich in dynamic models for understanding the cause and effects of outbreaks. Russell Reichelt developed a model to help determine the circumstances in which control programs were likely to be feasible. [53] Terrence Done reported on models he developed to understand questions about the long-term effects of repeated perturbations: How resilient were coral reefs? How long would recovery from predation take? How much predation, and how frequently did it have to occur before it caused local extinctions? [54]

Acanthaster and the Coral Reef appeared the same year (1990) as Birkeland and Lucas's book *Acanthaster planci: Major Management Problem of Coral Reefs*. Lucas and Birkeland agreed that predation had some potential for partially explaining the frequency and relative magnitudes of the outbreaks. However, it could not explain the times and places of outbreaks. Primary outbreaks occurred in particular years, with intervals in between. The mechanism was not provided as to why predators would be rare only in particular years. Therefore, they argued,

> With the information available at this time, the model that is most acceptable is one in which the potential timing and location of the outbreaks result largely from larval survival (strongly influenced by environmental conditions) and distribution by current patterns. Whether the outbreak occurs, and the magnitude of the outbreak when it occurs, can be controlled by factors influencing juveniles soon after metamorphosis, e.g., predation and disease. [55]

The difference in emphasis over the causes of the outbreaks reflected divergent research paths. That taken by Birkeland and Lucas followed the unique characteristics and natural history of *Acanthaster* as an individual species. In this regard they had polemics of their own about competing approaches to ecology. One of the major trends in ecology, they argued, was to focus on ecological processes and avoid referring to species and species-specific phenomena as much as possible. This was no exaggeration. For example, in the preface to his book, *Ecology of Coastal Waters: A Systems Approach* (1982), Kenneth Mann explained that he seriously considered writing the entire text without naming a single plant or animal. [56]

The crown-of-thorns provided vivid testimony that some species have such distinct, controlling effects on communities that they could not be simply averaged into any systemic approach that focused exclusively on energy transfer, trophic levels, or rate processes.[57] The crown-of-thorns outbreaks showed clearly how general and simplified models, which ignored the natural history and actual species composition of communities, had severe limitations for understanding ecological processes. "To examine the details of the natural history and biology of an individual species is presently looked upon with derision by some," Birkeland and Lucas wrote. "Models of population dynamics and predator–prey interactions are useful tools and have provided new insights into ecological processes, but they would not have predicted *a priori* the *A. planci* phenomenon by themselves."[58]

After all, about nine or ten species of other starfish in the Pacific occasionally feed on coral, but none had an effect like *Acanthaster*. To understand the *Acanthaster* phenomenon and its dramatic influence on coral-reef communities one had to appreciate the scale of its population fluctuations: outbreaks consisting of tens of thousands to millions of individuals, capable of devouring from 0.5 to 5 or 6 square kilometers of living coral in a single year.[59] One had to comprehend the unique natural history and qualitative characteristics of the organism: its pliable morphology, its large stomach surface to biomass ratio, its high fecundity of young adults, rapid growth of juveniles, early switching from juvenile to adult diet, high rate of food intake by adults.[60] The importance of recognizing the species that made up particular marine communities extended well beyond the *Acanthaster* phenomenon to fisheries in general. It implied that in order to develop adequate predictive regulatory procedures, each fishery had to be studied biologically in its own right.

Although the approach and polemics differed, those who emphasized predator–prey interactions and those who focused on the special characteristics of the organism converged on several fundamental issues. Anthropogenic disturbances had to be understood together with long-term patterns of ecological variation driven by natural forces. More than likely, crown-of-thorns outbreaks were part of the natural processes on coral reefs. However, the frequency and magnitude of outbreaks may be due to unprecedented human activities.[61] This meant, as Bradbury put it, "the outbreaks are neither particularly novel nor particularly natural."[62] "Natural" and "anthropogenic" were analytic distinctions; reality was not so easily cut and dried. As Birkeland and Lucas commented, "Mutually exclusive categorization should not imply mutually exclusive operations."[63]

It was possible that irreversible degradation would occur. Control measures were necessary, but not simply the kind usually called for—hand-collecting on individual reefs of commercial importance. Focusing controls on the effects of the problem—the starfish populations themselves—was not the solution. The roots of the problem had to be addressed. On Birkeland and Lucas's hypothesis, this would involve careful management of development along tropical coasts. Bradbury advocated controls over the suite of lethrinid-like fish predators of *Acanthaster* that had their density reduced on the Great Barrier Reef by commercial fishing over the previous twenty years. He further called into question the multiple-use management ethos of the Great Barrier Reef Marine Park, which, as he saw it, was underpinned by the assumption that the reef was mostly pristine, with only local degradation, largely stable, and capable of multiple use in different localities. This view, he argued, did not sit well with the evidence that the Great Barrier Reef "is largely exploited, with system-wide degradation, rapidly changing, and incapable of multiple use." [64]

The starfish populations declined on the Great Barrier Reef by 1990. A new outbreak began to appear six years later, while some of the reefs in Micronesia, the Ryukyus, and elsewhere remain in chronic conditions. The cause of *Acanthaster* outbreaks, whether natural or anthropogenic, remained unclear in the global coral-reef science community. It was still considered one of the most intriguing and urgent problems in reef ecology.[65] However, once at the center of concern for so many coral-reef ecologists, the crown-of-thorns now shared the spotlight with many other crucial issues demanding attention. Coral-reef scientists throughout the world recognized that coral-reef ecosystems were declining from the direct and indirect disturbances of relentlessly expanding human populations along with commercial and industrial developments in tropical coastal areas. At the same time, a new factor entered the scene: global climate change. All these issues demanded immediate attention. Evidence for the reality of such changes, their causes, and their effects captured the attention of coral-reef scientists in many countries. The long and heated controversy over the crown-of-thorns offered a precedent for what they should be thinking about and a measure for how they ought to act. As they grappled with such issues, they came to see that they had to address, in a coordinated manner, the problem of distinguishing between anthropogenic changes and long-term natural oscillations.

13

Coral Bleaching and
Global Warming

Although it appears that elevated local seawater temperatures caused bleaching, linking this effect to global warming cannot be conclusive at this time ... Several international monitoring efforts are now in progress or are planned so that the appropriate data can be gathered.

Barbara Brown and John Ogden, 1993

Twenty years ago many coral reef scientists overreacted to the crown-of-thorns problem, but this time with the bleaching issue something quite different seems to be happening.

Richard Grigg, 1992

During the 1990s, international conservation agencies and governments of many countries came to recognize the dangers facing coral reefs. A survey conducted on behalf of the International Union for the Conservation of Nature (IUCN) in 1984–89 indicated that people had significantly damaged or destroyed reefs in ninety-three countries.[1] Coral reefs covered some 600,000 km^2 throughout the tropical world, and there were estimates that about 10% of them were degraded beyond recovery, with another 30% likely to decline within the next two decades. Reefs identified as being at greatest risk are in South and Southeast Asia, East Africa, and the Caribbean.[2] Some of the most serious problems are over-fishing and destructive fishing methods, including the use of dynamite and sodium cyanide; sedimentation and pollution; rapid population growth in tropical developing nations; and tourist-related activities and associated impacts.[3] To some, the degradation of reef ecosystems seemed practically irreversible because it appeared impossible in practice to put an end to the human impact.[4]

Ecologists generally were, and remain, less certain of the cause of other disturbances on coral reefs. Some categorize the crown-of-thorns plagues as "natural stresses," along with storm damage, earthquakes, and wave action.[5] Others are unsure whether such outbreaks were natural or result from human interference and, in particular, the increased load of nutrients to reef systems.[6] The uncertainty over the crown-of-thorns in the 1990s was often compared to another newly observed global environmental perturbation: coral bleaching. There was no question that bleaching was real and that it caused serious destruction of coral reefs. The major controversy was over the scope and significance of the phenomenon—whether it provides the first indicator of ocean warming resulting from global warming.

The main issues were indeed similar to those that framed the crown-of-thorns controversy for so many years: Was widespread coral-reef bleaching truly a recent phenomenon, related to anthropogenic impacts? Or was it an artifact of reporting, a result of more observers in recent years with a heightened awareness of environmental impacts? Was it really caused by elevated sea-surface temperatures and related effects? What are the effects of bleaching on coral-reef communities? How resilient are coral reefs? How soon would recovery occur? How much bleaching must occur, and how frequently did it have to be repeated before it caused local extinctions?[7]

Although the issues were similar, the response of the coral-reef science community differed dramatically from that in the early controversy surrounding the crown-of-thorns. Those differences highlight gross changes in the institutional growth of coral-reef environmental science and coral-reef management, the political importance afforded to global environmental issues, and coral-reef scientists' hard-won awareness of the general need for baseline data from which to distinguish between human-induced changes and long-term natural processes.

Breaking Symbiosis

Understanding coral bleaching begins with appreciating the physical constraints on the intimate symbiosis of algae living in the transparent tissue of coral polyps. Between one and two million algae cells per square centimeter of coral tissue give corals their splendid colors and help to nourish them with their photosynthetically produced carbon compounds. Some species are known to receive 60% of their food in this way. Algal photosynthesis also accelerates the growth of the coral skeleton by causing more calcium carbonate to be produced. The association enables algae to obtain compounds that are scarce in the nutrient-poor waters of the tropics, where warm surface waters overlie and usually lock in cold, nutrient-rich waters.[8]

When corals bleach, the delicate balance between algae and coral polyp is destroyed. Corals lose algae, leaving their tissues so transparent that only the white calcium carbonate skeleton is apparent. Other organisms such as anemones, sea whips, and sponges—all of which harbor algae in their tissue—are also known to whiten in the same way. Some of this is recognized as routine. When subject to sporadic adverse changes, such as temperature increases, corals release increased numbers of algae.[9] Unless the algae return, the coral polyps starve to death. Even small temperature increases above the normal local maximum temperature may result in bleaching. It can occur in summer months when the average temperature exceeds the seasonal average by 1°C. Most corals survive infrequent bleaching episodes, but repeated or sustained bleaching events kill them.[10]

Laboratory studies of the 1980s demonstrated that other environmental stresses such as increased ultraviolet light, sedimentation, and toxic chemicals, as well as cold ocean temperatures, can also cause bleaching. Coral reefs are found in oceans where the annual temperature range varies from 20 to 30°C. Thermal tolerance seems to vary by coral species as well as geographic location. Reefs in the Florida Keys grow at 18°C; temperatures above 33°C are tolerated by healthy coral communities in the northern Great Barrier Reef and the Persian Gulf.[11] In addition to the varying temperature sensitivity of corals, there may also be varying sensitivities of diverse varieties of algae within and among host species at different depths.[12] Coral bleaching was first reported by Thomas Goreau in 1964 off the south coast of Jamaica,[13] but there were few reports of it anywhere until the 1970s.[14] Bleaching associated with abnormally high seawater temperatures, or prolonged peak sea-water temperatures in the summer, became a frequent problem in the 1980s and 1990, especially in the eastern Pacific and the Caribbean.[15]

El Niño

Periods of unusually warm waters off the equatorial coast of South America, along with their related effects, had been known for centuries by peoples of the equatorial Pacific. Such events were called El Niño (the Christ Child) by Peruvian fishermen because these effects tended to peak around Christmas time. Oceanographers of the 1980s recognized El Niño events as one phase in a quasi-periodic two- to seven-year cycle resulting from interaction between atmospheric circulation and the dynamics of the layer of warm, light water resting on a much deeper layer of cold water.[16] In one phase (the neutral or cold phase) of the cycle, trade winds blow from east to west, pushing water away from the South American coast. As a result, the ocean's surface is about 60 cm (2 ft) lower off Peru than it is off Indonesia. To replace the water that the wind has pushed away, cold nutrient-rich water from the depths wells up as a sort of pumping action. When an El Niño begins, the patterns of wind and upwellings change. The trade winds dwindle or begin to blow from the west. Peru had the largest fish catches in the world from 1962 until 1972, when the arrival of the warm El Niño caused anchovies and other fish to move elsewhere in search of food. As the warm water spread into the central and eastern Pacific, the reverberations were felt far beyond the collapsed fisheries of Peru, resulting in drought and crop failures in many areas around the world. As one of the world's great weather makers, El Niño's effects are variable and may cause droughts in Brazil, southern Africa, and Australia, and flooding from California to Cuba.

Many times during a ten-month period in 1982–83, a remarkably strong El Niño Southern Oscillation (ENSO) event warmed the waters of the eastern Pacific 3° to 4°C over the seasonal average. Oceanographers described it as the strongest warming of the equatorial Pacific this century, perhaps in one hundred or more years.[17] It left two thousand people dead and $13 billion in economic losses. Interest in El Niño on the part of policy-makers and scientists rose sharply.

The dramatic and devastating effects of the 1982–83 El Niño on coral reefs were first documented by Peter Glynn in November 1983.[18] He was in the process of moving from the Smithsonian Tropical Research Institute in Panamá to take up a position at the University of Miami. Before leaving, he checked the study sites that he had been monitoring for fifteen years. He was about to conclude another year of sampling—for species interactions, crown-of-thorns predation, abundance of coral in different zones, fish abundances—when he saw that his study reefs were dying. He tracked the bleaching event on reefs off Panamá, Costa Rica, and the Galápagos Islands. As a result of the elevated temperatures, between 70 and 90% of the corals in Panamá and Costa Rica perished as well as more than 95% of the corals in the Galápagos. Four years later, Glynn reported that all three areas showed minimal signs of recovery.[19] Similar mass mortality was observed at Colombia and Ecuador.[20]

Coral bleaching was reported at many other places during the

1982–83 El Niño, including parts of the Caribbean, the Society Islands, the Great Barrier Reef, the western Indian Ocean, and Indonesia. It resulted in loss of as much as 90% of the coral cover on the shallow reefs of the Thousand Islands in the Java Sea. Five years later coral cover was still only 50% of its former level.[21] Another ENSO event in 1987 brought reports of mass bleaching throughout the Red Sea, the Caribbean, and Bermuda.[22] Coral-reef bleaching occurred again in the Caribbean in 1989 with severe effects up to November 1990.[23]

Isolated instances of bleaching were known to commonly occur in response to heavy rains, pollutants, decreased salinity, or other local stresses. But never before, it seemed, had bleaching occurred so frequently and almost simultaneously across such wide swaths of the tropics.[24] Did this remarkable increase provide empirical evidence for global warming? Were the frequency and intensity of ENSO events increasing because of global warming? Was sea-surface temperature rising because of enhanced greenhouse effects?

CANARY IN THE COAL MINE

Although the term "greenhouse effect" is often confused with the controversial issue of anthropogenic global warming, the heat-trapping properties of the atmosphere, its gases, and particles are well understood and well validated. It is as good a theory as there is in science. It explains, for example, the very hot conditions under the thick atmosphere of Venus and the very cold conditions under the thin, weak "greenhouse" of Mars.[25] What is controversy-laden, is global warming due to the burning of fossil fuels, higher concentrations of atmospheric CO_2, and other contributors to greenhouse gases: chlorofluorocarbons (CFCs), methane, and nitrous oxide.

In 1988, the United Nations General Assembly set up an International Panel on Climate Change (IPPC) to advise world leaders on the seriousness of the problem. In 1990, three hundred climate scientists from more than twenty countries concluded that a warming of some 0.5°C had occurred globally over the past one hundred years.[26] Based on computer simulations of climate, they predicted that if greenhouse-gas emissions continue at their present rates there would be a doubling of atmospheric CO_2 by the middle of the twenty-first century. This would result in a mean global temperature increase of about 1°C, and a warming of up to 2°C in the tropics. This in turn would lead to a rise in sea level of about 5 cm per decade. Global warming would also entail more frequent and extreme weather conditions. There would also be depletion of the ozone in the stratosphere as a result of chlorofluorocarbon emission, resulting in an increase in the earth's exposure to ultraviolet light.[27]

A Global Task Team sponsored by UNEP (United Nations Environment Program), the IOC (Intergovernmental Oceanographic Commission) of UNESCO (United Nations Educational, Scientific and Cultural Organization), and the Association of South Pacific Environmental Institutions examined the prospects for coral reefs. Working together with the World Meteorological Organization (WMO) and the World Conservation Union (ACNE), the Task Team warned, "the condition of many of the world's coral reefs has reached a crisis point. . . . Global climate change may directly impose new stresses on reefs, or it may interact synergistically with other more direct human pressures to cause added and accelerated environmental damage." [28] It would not pose an immediate threat to the existence of reefs worldwide, but the impact locally would vary from disastrous to benign. Some reefs might be destroyed, others badly affected, while some could even experience more vigorous growth in the medium term.[29]

Although environmental organizations and government leaders were concerned with making preventative policy, many critics dismissed global warming as merely media hype, arguing that "apocalypse sells." They feared that the "greenhouse effect" would push governments into a costly mistake that would stifle the "free-market economy," and that increased government encroachment on individual freedom would result in a new kind of "social control." [30] Scientific opinion varied and continues to differ as much as it did in the early debates over crown-of-thorns outbreaks. Some accept the evidence for a general warming trend over the past hundred years, but doubt that it is necessarily due to human activities and increasing greenhouse gases. Other natural factors ranging from volcanoes to the sun-spot cycle might be involved.[31]

Many others denied the validity of any evidence for global warming. They pointed to the unreliability of temperature records for the past century and to the uncertainties involved in climate predictions based on mathematical models and computerized simulations.[32] They also argued that the time frame chosen determines the answer: although there was an increase from 1880 to 1940, from 1940 until 1960 the temperature dropped so much as to entail predictions of a coming ice age.[33] Only one issue was certain: more data were needed.

Could coral reefs serve as the canary in the coal mine—the first indication of global ocean-temperature increases? The spectacular and sudden bleaching, especially in the Caribbean in 1987, led to much speculation in the press, testimony in Congress, and intense discussion among coral-reef researchers.[34] Special bleaching sessions were organized at scientific meetings in Mayagüez, Puerto Rico, Curaçao, Sarasota, Florida Keys, San Salvador, Bahamas, Townsville, and Havana, Cuba.[35] A special volume of *Coral Reefs* was dedicated to the problem in 1989.[36] Two hearings were held in the U.S. Senate addressing the bleaching issue, one in 1987, the other in 1990.[37] The Senate required the National Science Foundation to develop a plan by May 1991 to address research on bleaching and reefs in the context of global change.[38]

The cause of the bleaching events and their association with putative global warming were assessed from diverse perspectives and received with varying degrees of skepticism and agreement. The favored interpretation was that they were due to elevated sea temperatures in combination with changes in solar irradiance (especially ultraviolet light): either decreased light penetration due to sedimentation or increased light from calm, almost doldrum-like water conditions.[39] But an array of viewpoints existed.[40] While some saw bleaching as an early signal of global warming, others considered it an indicator of stresses on reefs caused by local human activities or the natural variability of the environment.

Thomas Goreau, Jr., and Raymond Hayes testified in Congress that coral bleaching throughout the Caribbean was due to increased ocean temperatures and that it was possibly a harbinger of global warming.[41] They warned that repetitions of such events could alter the food chain of the coral-reef ecosystem, making corals less able to compete with rapid growth of fleshy algae. Bleaching could ultimately result in "severe economic losses from deterioration of reef fisheries, tourism and shore protection." [42] Others agreed that coral-reef bleaching might well be a harbinger of global climate change, but insisted that "firm conclusions would be premature at this point." [43] Still others were more certain.

In 1990, Ernest Williams Jr. and Lucy Bunkley-Williams at the University of Puerto Rico brought scattered reports together to document the occurrence of worldwide coral-reef bleaching events in 1979–80, 1982–83, and 1986–88. It was abundantly clear to them that they were caused by "the general global warming trend in the 1980s." [44] Their predictions were dire: "At the very least, reef ecologies will be altered. At the very worst, the coral reefs, which have been able to adapt to gradually changing conditions in the geologic past, may not be able to cope with more rapid climatic changes associated with the greenhouse effect and may perish altogether." [45] They warned that the cycle may repeat in

1991 and 1992, "possibly with more intensity, and will probably con-tinue and increase until coral-dominated reefs no longer exist."[46] Coral reefs, they asserted, were not just the victims of global warming, they were also part of its solution. They were as important as tropical rain-forests in reducing greenhouse gases. As corals deposit calcium carbon-ate for their skeletons, they remove a large volume of CO_2 from the oceans. "Ironically, damage to this undersea ecosystem could accelerate the very process that hastens its demise."[47]

The portrayal of reefs as major sinks for CO_2—and therefore poten-tial saviors from increases in greenhouse gases—received considerable public attention. But it did not go without contradiction. Donald Kinsey at the GBRMPA asserted, in 1992, that "in fact, they presently act as a sink for the equivalent of only about 2% of the anthropogenic CO_2."[48] At the same time, critics pointed to a lack of high-quality tem-perature observations in the Caribbean and elsewhere to argue that even the evidence that temperature had caused some of the bleaching events was inconclusive.

CALL FOR MONITORING

In June 1991, with the support of the National Science Foundation, NOAA (National Oceanic and Atmospheric Adminstration), and the Environmental Protection Agency, fifty reef scientists and climatolo-gists convened in Miami to discuss the bleaching problem, coral reefs, and global climatic change. The meeting was held in direct response to the U.S. Senate hearings. As the organizing committee of the bleaching workshop wrote,

> The view of several panelists who testified was that elevated sea-surface
> temperatures (SSTs) are likely to be the cause of the bleaching, and that
> the phenomenon was the harbinger of global warming. Wary heads of
> federal agencies represented at the hearings have sensed the need to cover
> their bases—not only for political reasons, but also because of concern
> about the scientific basis for legitimately viewing the phenomenon as an
> indicator. Accordingly, the topic of coral bleaching as an indicator of
> global climate change has particular currency in Washington.[49]

All scientists at the Miami workshop agreed that since intensive un-derwater observations on coral reefs began in the 1960s, there was evi-dence that the frequency and scale of bleaching had increased within the previous ten years. There was also paleoecological evidence that ex-treme bleaching and mass mortality comparable to the 1982–83 event had not occurred in the eastern Pacific during the previous three hun-dred years.[50] Nonetheless, one could not conclude that this indicated the advent of global warming—for two reasons. First, it was not at all clear that all reports of bleaching were actually due to increased temperature. The evidence for such a correlation was strong for bleaching events in the eastern Pacific. In other cases, it seemed likely that factors such as irradiance, reduced salinity, high sedimentation, and other human-induced changes played a role.[51]

Second, and more importantly, bleaching observations had not been made on the temporal and spatial scales necessary to make meaningful predictions about climate change. There was simply a lack of long-term, ecological data on climate change.[52] For the purpose of prediction, climatologists defined "climate" in terms of "quantitative measurements of primary physical and chemical variables over multidecade and global or large-region averages, with associated statistics."[53] A decade of regional observations of coral bleaching fell far short of the time and space scales necessary to define or detect a climate trend.[54] All participants at the Miami workshop agreed that there was "no clear evidence for the coupling of coral bleaching and global warming because of the lack of relevant data."[55]

They not only lacked the hard evidence on global and local bleaching trends to satisfy the standards of climatologists in regard to global warming, they also had no historical data about coral degradation caused by other factors. Most believed, intuitively, that coral reefs were in immediate danger in many areas due to increasing agriculture, deforestation, and urban development. They thought that the direct effects of the human population explosion, resource exploitation, and development outweighed the future threat of climate change. Their "primary conclusion" was that "on a global average basis, coral reefs are being lost or degraded at an alarming rate."[56] Again, they simply lacked the data needed to confirm, quantify, or explain this trend "on a scientific basis."

They also lacked the models needed to understand the long-term effects of environmental perturbations.[57] Joseph Connell emphasized how coral-reef scientists tend to favor healthy reefs as their study sites, thus biasing subsequent observations.[58] Other aspects of their knowledge were colored by the relatively short period of "the scuba age of underwater observation" relative to the constants of natural variability in development of reef structures.[59] To tease out and distinguish human-induced impacts from natural perturbations, they needed to retrospectively extend their baseline data on change, variability, and biological responses in coral-reef environments. Such environmental signals could be obtained from studies of the growth of coral skeletons.[60] Some corals showed annual florescent bands that could be read like tree rings to obtain a picture of past climate and environmental conditions on a particular reef. The bands in corals growing on the shore side of a reef, for example, are caused by fulvic acid leached from soil by the water; they thus provide a record of nearby river output, and also rainfall. Because massive corals may be many centuries old, they thus contain a long-term record of floods, droughts, and river flow—on which climate modelers can work. The Miami workshop also "strongly and unanimously recommended development of a global-scale, coordinated program of coral-reef monitoring." [61] Five to ten intensive stations and twenty or so other sites around the world would be required to serve such a long-term monitoring program. In short, the idea was to set up a "coral watch" program.[62]

Heralded throughout the coral-reef community, the Miami workshop's conclusions about coral bleaching were reported with diverse interpretations. One news item in *Science* declared that "the National Science Foundation-sponsored meeting of reef scientists concluded in 1991 that global warming was not the culprit." [63] But this was a misleading statement. The real issue was that appropriate data were lacking. Many leading coral-reef scientists remained circumspect, including Barbara Brown, co-founder of the International Society for Reef Studies and director of the Center for Tropical Coastal Management, and John Ogden, director of the Florida Institute of Oceanography. When referring to the workshop, they wrote: "the paucity of knowledge about the physiological response of corals to stress and temperature, the inadequacy of seawater temperature records and the lack of standardized protocol for field studies made it impossible to decide whether bleaching reflects global climate change in the ocean." [64] Bleaching, together with the toll taken by pollution and overfishing, they warned, would seriously burden the future economies of many developing nations:

> If the temperature increase of one or two degrees C, predicted by the Intergovernmental Panel on Climate Change, does take place over the next fifty years in the tropical latitudes, the consequences for coral reefs could be disastrous. Unlike the miners with the canary, we cannot yet link bleaching to a clear cause. But that does not mean we should ignore coral's message.[65]

Glynn also kept an open mind about the relation between coral bleaching and global warming. As he wrote in 1991,

> Whether or not the coral bleaching complexes of the 1980s have been caused by global warming is a critically important yet currently unresolved question. However, the evidence for numerous recurrent,

large-scale coral bleaching and mortality events in recent years is un-equivocal. Like the destruction of tropical rain forests, the world's coral-reef ecosystems are facing major disruptions. Indeed the deteriora-tion of some coral reefs has been caused by increased sediment and eutrophication associated with rain forest destruction.[66]

Questions of global warming and coral bleaching were of vital con-cern to coral-reef scientists worldwide. At the seventeenth Pacific Sci-ence Congress held in Honolulu in 1991, several papers addressed how sea-level change and sea-surface temperature increases would affect coral reefs. If global warming leads to an increase in the magnitude and frequency of ENSO events, some warned, coral reefs might be threat-ened with extinction.[67] But, as in the Miami workshop, the most diffi-cult challenge to emerge from the symposium in Honolulu was the problem of separating anthropogenic change from natural variability, and developing successful management strategies. Again, coral-reef sci-entists called for long-term monitoring studies in a variety of reef set-tings to establish baselines against which anthropogenic impacts could be measured.

Establishing Baselines

Monitoring was a dirty word; it was not question-driven research; it was considered "mindless"—before coral bleaching and global warming forced itself on the political agenda.[68] Although biologists linked coral bleaching to higher ocean temperatures, it became abundantly clear that they did not even know, for example, what the temperatures in Ca-ribbean waters were on an annual basis. To understand the cause and ef-fects of environmental perturbations, and to manage coral reefs effectively, one had to operate on the same geographical and temporal scales nature is operating on. The scientific and management potential of monitoring became widely appreciated during the 1990s. Various re-gional and international monitoring networks were organized.

One of the first was the Caribbean Coastal Marine Productivity net-work (CARICOMP). Set up in 1990, it was a cooperative research net-work of more than twenty Caribbean marine laboratories, parks, and reserves in sixteen countries.[69] It began systematic observations in 1992, establishing research sites, mapping the distribution of coral reefs, sea grasses, and mangroves, and collecting monitoring data on productivity using standardized techniques. The network's goals were to determine the dominant influences on coastal productivity, discriminate human disturbance from long-term natural variation, and distinguish between local perturbations and region-wide changes. It aimed to use a series of protocols to make the same observations, on the same schedule. By comparing one research site to another over a long period, one would be able to correlate observed differences and similarities (the population density of certain fish, for example) with particular factors.

There were ample illustrations that the Caribbean region had systemic properties operating over large spatial and temporal scales.[70] The mass mortality of 90 to 99% of the black spiny sea urchin *Diadema antillarum* in 1983–84, presumably caused by a pathogen, followed well-defined tracks of ocean currents throughout the Caribbean. *Diadema* was acknowledged as a keystone species whose grazing on algae prevented it from outcompeting, overgrowing, and killing corals.[71] *Diadema* populations had failed to recover even ten years later, despite their high fecundity.[72] White-band disease, resulting in decades-long dieback of *Acropora* in the Caribbean, indicated a pathogenic condition operating on a much longer timescale.[73] Coral bleaching, of course, also had region-wide impact. Coordinated by a Data Management Center at the University of the West Indies, the CARICOMP network would be able to respond rapidly to coral bleaching, diseases, and periodic oceanographic phenomena.[74] A similar network of marine laboratories was planned for the central and western Pacific Ocean.[75]

Obtaining funds for the CARICOMP network as a whole proved difficult. As John Ogden, co-chair of CARICOMP's steering committee, recalls, the attitude among funding agencies such as USAID (Agency for International Development) was that such regional programs did not work because the participating laboratories were in such culturally diverse countries, with different languages, a different sense of what is important, and different concerns about scientific imperialism. However, CARICOMP grew out of the Association of Marine Laboratories of the Caribbean, which had held annual meetings since the 1960s. That association had been a means for cutting through such differences. As Ogden later saw it, the success of CARICOMP was in fact due to the commmonality of culture among the scientists involved:

> The sociology of scientists is identical. It doesn't matter if you are Spanish-speaking or whatever—marine labs are marine labs wherever they are, and they have a certain structure, and they're all the same. So you are right at home when you are at one. The problems they have in surviving, dealing with their administration, are all the same at different locations. So you have this kind of warm and fuzzy feeling with those people. So everyone understands their problems and tries to help each other out.[76]

There was still another issue. Funding agencies wanted bilateral arrangements. For example, the United States supporting Jamaica, the United States supporting Barbados. Moreover, CARICOMP included Cuba, which as far as American funding agencies were concerned had to be whited out.[77] Ogden and his colleagues recognized that obtaining and maintaining central funding for CARICOMP was simply not feasible. Instead, the group aimed to work with members at each individual site to provide them with the best possible arguments for their institutions and agencies for supplying funds. UNESCO provided funds for annual meetings, as well as support for its data-management center. Issues of property rights over data also had to be addressed. How long did the data CARICOMP produced belong to the network before it belonged to everyone? The group agreed that each laboratory could do what it wanted with its own data. But the network was to have exclusive rights to the pooled regional data for one year, after which they were made available to the coral-reef community.[78]

In 1992, at the Seventh International Coral Reef Symposium in Guam, a Global Coral-Reef Monitoring Network was endorsed as part of the Global Ocean Observing System (GOOS). It was subsequently established as a co-sponsored initiative of the Intergovernment Oceanographic Commission (IOC), United Nations Environmental Program (UNEP), and the World Conservation Union (ACNE). It aimed at monitoring selected worldwide sites for coastal and near-shore phenomena related to climate change and all other perturbations. It also planned to strengthen regional networks of institutions by providing facilities for interregional and global collaboration in data handling and access.[79] During its first years of operation it was situated at the Australian Institute of Marine Science and at the International Center for Aquatic Resources Management, Manila.

In 1994, the International Coral Reef Initiative (ICRI) was born to provide a focus on the plight of reefs and actions necessary to reverse the trend of degradation. It was supported by the governments of Australia, France, Jamaica, Japan, the Philippines, Sweden, the United Kingdom, and the United States. A framework for action was developed as a basis for achieving the sustainable use of coral reefs and associated ecosystems.[80] Although many countries had established natural reserves and national parks to preserve coral-reef ecosystems, many others lacked them, as well as personnel with sufficient training. The ICRI held regional and international workshops to provide forums for nations to determine conservation and management strategies and foster communication between scientists and managers.

The ICRI also focused on the coordination of international, national, and regional research and monitoring programs. Such monitoring networks would provide policy makers and environmental managers with information on local and regional trends in demographic, economic, and environmental conditions. They were also directed toward answering questions about specific issues such as fisheries, pollution, and tourism, and to help discriminate between natural variability and human impact. The complex ways in which coral reefs and human societies interact meant that there was no single approach for managing reef resources for their conservation. Managers of marine parks and reserves were encouraged to work together with scientists. They were also encouraged to listen to indigenous people and make use of the knowledge and expertise of local communities. By 1996, seventy-five countries had participated in ICRI workshops at global or regional levels.[81]

14

CASSANDRA AND THE SEASTAR

"Oh, Pangloss!" cried Candide. "A scandal like this never occurred to you! But it's the truth, and I shall have to renounce that optimism of yours in the end."

"What is optimism?" asked Cacambo.

"It's the passion for maintaining that all is right when all goes wrong with us," replied Candide.

Voltaire, *Candide*

Coral-reef environmental science has come a long way since the days of "Save the Barrier Reef" bumper stickers in Australia, and the Westinghouse survey in Guam. The International Coral Reef Initiative would encourage and help governments and international organizations to strengthen their commitment to conserving, restoring, and promoting sustainable use of coral reefs and associated environments. Networks of laboratories, sharing and comparing data, would help resolve complex problems of distinguishing between human-induced impacts and long-term natural oscillations.

Coral-reef scientists, ever so self-conscious, also continue to monitor their own behavior. How ought they behave in the midst of global environmental uncertainties? How does the manner in which they address climate change and coral bleaching compare to thirty years earlier when *Acanthaster* outbreaks were first reported? What, if anything, had they learned from the crown-of-thorns controversy? What should they be thinking about in regard to other global environmental uncertainties? At the University of Hawaii, Richard Grigg offered an assessment of these issues in a plenary lecture for the annual meeting of the International Society for Reef Studies in 1991.[1]

Grigg began with a discussion of the Greek myth of Cassandra, the princess of the ancient city of Troy. It is said that the god Apollo loved her and promised to give her the gift of prophesy in return for her affections. Cassandra accepted the proposal and then spurned Apollo's advances. In revenge, Apollo cursed her so her prophesies would never be believed. When Cassandra predicted disaster if a large wooden horse left outside the besieged city was brought in, the Trojans ignored her warning. The wooden horse was dragged into the city where, once inside, a group of Greek soldiers hidden in the huge horse emerged. The city was destroyed.

Cassandras today are usually viewed as "prophets of doom." Some are real, Grigg asserted, and scientists should take them seriously since they "base their views on the best scientific information available." But there are also "false Cassandras" whom they *should* be wary of, "Those who tell us 'the sky is falling' when in fact it is not." [2] Ironically, they are often believed, particularly by the public at large. As an example, Grigg quoted Jacques-Yves Cousteau, who in 1971 stated that "in 10 years there would not be any fish remaining to take out of the sea." [3] Perhaps, he thought, most coral-reef scientists might grant Cousteau some degree of "poetic license" for this kind of exaggeration. But how much poetic license should they themselves be granted? Grigg turned to the crown-of-thorns controversy and pointed out Richard Chesher as one of the earliest and best-known "false Cassandras." He quoted Chesher's statement in 1969: "There is the possibility that we are witnessing the initial phases of extinction of Madroporarian corals in the Pacific." For another over-reaction, he cited a "State of the Earth" report from the Smithsonian Institution for the Center for Short-Lived Phenomena asserting that "if the starfish explosion continued unchecked, the result would be a disaster unparalleled in the history of mankind." [4]

All these predictions were false, Grigg asserted. Certainly, many reefs had undergone serious damage, but there was not a single species of coral that had become extinct, nor had any reefs of significant size disappeared as a result of starfish predation. In 1970, there had been widespread concern that the outbreaks were caused by anthropogenic factors. But today, he declared, "Most scientists agree that population fluctuations are more likely due to natural year-class variations caused by high fecundity and variable recruitment success." [6] He insisted that it was now well established that population oscillations of *Acanthaster* have been going on for at least 7000 years. He pointed to recent geological studies showing a pattern of abundance of *Acanthaster* spines recovered from sediments laid down during the Holocene transgression on the Great Barrier Reef.[7] Given the evolutionary age of the starfish,[8] Grigg suspected the cycle was probably millions of years old.

Grigg admitted that his description was somewhat oversimplified. The sediment record didn't speak for itself. He was confident, however, that interpretative uncertainties and taphonomic problems would be resolved with studies of less disturbed sediment records. He also recognized that anthropogenic factors, especially nutrient runoff from coastal developments, may have affected the outcomes of some crown-of-thorns outbreaks. Nevertheless, he argued, coral reefs in the Pacific were not threatened by extinction, and nature would continue to play a prominent role in controlling starfish abundance. Interestingly enough, he commented, this outcome was predicted in 1972 by Thomas Dana, William Newman, and Edward Fager. But few believed them. Grigg lamented, "It seems that the false Cassandras of the day had generated so much publicity that the correct message was barely heard."[9]

Yet there was an optimistic ending to Grigg's tale. Whereas many coral reef scientists had overreacted to the crown-of-thorns problem, something quite different was happening with the coral-bleaching issue. Sure, there were some who had testified in Congress and claimed that coral bleaching may possibly be a harbinger of anthropogenic global warming. But Grigg pointed to the Miami workshop, Coral Bleaching, Coral Reef Ecosystems and Global Change, as a measure of just how mature coral-reef science had become. "This time the coral-reef scientific community organized themselves in a timely and responsible manner and came forward with a reasoned response based on the best information available." They also came forward with recommendations for future action.

Now, "the Cassandras will be tested by the truth of careful experimentation, long-term monitoring, and objective interpretation. ...This time little poetic license was granted to the false Cassandras and the real story is being heard." Grigg himself doubted that human-induced global warming was occurring. He believed that long-term fluctuations in global temperature were correlated with the sunspot cycle. It was clear to him, however, that greenhouse gases were increasing in the atmosphere and would increasingly affect the natural variability of the global climate system. But he thought it unlikely that "greenhouse will become a death-house for coral reefs."[10] Generalizing the problem of overreaction to environmental issues, he noted that all of environmental science seemed to be affected by doomsday thinking.

> It is as though the four horsemen of the apocalypse, as described in the Bible in the Book of Revelation to John the apostle, continue to haunt the psyche of humankind: war, hunger, pestilence, and death. Perhaps, it is our fear of the inevitability of death that perpetuates doomsday thinking. It is a theme that appears over and over again in the cultural history of mankind.[11]

Many commentators have called on millenarianism when discussing environmental "overreaction" and doomsday thinking in the late twentieth century. They point to the waning years of the tenth century when millions braced themselves for the Apocalypse, believing that the approaching year 1000 was the very millennium—the end of the heavens and earth prophesied in the Bible's Book of Revelation. The prospect of the impending Day of Wrath terrified people into rash and foolish actions. Some gave away all their possessions; others hastened to do harsh penance for their deeds.[12] In this final decade of the twentieth century, the hand of God is usually replaced by more visible agents: belching smokestacks, gasoline-powered automobiles, power-generating stations, ferocious destruction of forests. There are also less visible agents at the top of the food chain, pathogenic bacteria and viruses.[13]

When explaining the popularity of apocalyptic writers and the motives of "the environmental movement," some writers stress that people choose what to fear. Some emphasize the Cold War as having had a dramatic effect on perceptions of the future by showing the real threat of global catastrophe.[14] Indeed, for many decades the threat of nuclear bombs has been associated with global environmental change, especially invasions, and population explosions of plants, animals, microbes, as well as of our own species. The analogies have survived the Cold War. When describing zebra mussels, natives of southern Russia, and their rapid infestation of North American waterways in the late 1980s as a result of the careless dumping of ballast water from a transatlantic freighter, biologists wrote in *BioScience*,

> Just when the threat of a Russian invasion of North America seemed to have disappeared with the end of the Cold War, an invasion has been found to be not only under way but proving to be successful. Rather than missiles, a naval force of hordes of zebra mussels has secured beach-heads in many US and Canadian lakes and rivers. One could call it biological warfare, but not directed by any human admiral.[15]

The end of the Cold War has also been explicitly discussed in official

American science and technology policy with the threats of global warming and emerging diseases. In 1995, John Gibbon, assistant to the President for Science and Technology, wrote of the politics of science in the United States:

> Fifty years ago, in his office at the Carnegie Institution on 16th Street here in Washington, D.C., Vannevar Bush was putting the finishing touches on a document that was to be the blueprint for U.S. science and technology for the rest of the century. Bush was writing for a world of change and transition, a world where the only clear enemy was Communism and the battle was waged in terms of technological superiority.
>
> We approach the twenth-first century in a similar period of sweeping change, although we have different enemies than those Bush envisioned. They are not armies; they are new and emerging diseases. They are not missiles, but the threat of rapid global change. They are not tanks and submarines; they are poverty, crime, and economic stagnation.[16]

Yet critics of environmental policy raise many nonscientific reasons to explain our "unwarranted obsession" with environmental problems. Some point to the Judeo-Christian heritage—and the idea of "original sin," the fall from Eden—to understand the roots of the environmental movement.[17] Others invoke environmentalists' rejection of the biblical injunction for humankind to be fruitful and multiply and hold dominion over other creatures and Earth.[18] Instead, critics try to reveal hidden values of socialism and Ludditism that threaten to undo the scientific and industrial revolution: "The prophets of apocalyptic doom mislead and scare the public with their warnings of impending catastrophes."[19] Some writers predict that we will soon enter a new optimistic age of "ecorealism." It would be based on the premise that "logic, not sentiment, is the best tool for safeguarding nature," that we must learn to "think like nature," and recognize that nature is an "elaborately defended fortress" that has been repelling more serious assaults for four billion years.[20]

Grigg, however, was not an environmental Pollyanna. As we face the twenty-first century, he warned, such a doomsday scenario as described in the Bible may come true. He pointed to what Garrett Hardin has called "the population taboo."[21] With the human population expected to reach six billion by the year 2000, our species has every reason to worry. In Grigg's view, the worldwide decline of coral reefs may have more to do with human urbanization than with an enhanced greenhouse effect. "The real truth may be that 'spoken' by the reefs and that for either one of us to survive, mankind must discover a way to control human population growth."[22]

Today, evidence for a relation between coral-reef bleaching and prolonged high ocean temperatures is strengthening, and many researchers continue to suspect global climate change. Whatever merit we may attribute to Grigg's views on the global decline of coral reefs, little can be granted to his views of the crown-of-thorns controversy and the lessons he draws from it about how scientific objectivity was effectively drowned out by the cries of alarmists. This view invites criticism from various perspectives. Let's start with Grigg's claim that Dana, Newman, and Fager heralded "the correct message" that went unheeded. We should remember that the Scripps biologists denied that there were any real population explosions of the crown-of-thorns. Their model was based on the starfish's behavioral characteristics: dense aggregations formed in search for food following typhoons. Certainly the director of Scripps, William Nierenberg, believed the controversy had been resolved by his colleagues when he attempted to use it to dismiss the whole issue as a sort of "hoax." For him, the crown-of-thorns was simply a good illustration of the false problems conjured up by environmentalists and the media, and which only threatened to distract good scientists and push science itself off its true course. He had also misinterpreted their theory, which was seriously scrutinized and subsequently rejected by all crown-of-thorns investigators.

Therefore, to accept the views of Dana, Newman, and Fager in the midst of the controversy would be to disregard the very scientific criteria—objectivity and detached scrutiny—that Grigg called for in making such assessments. Insofar as the biologists at Scripps denied any real outbreaks of the crown-of-thorns, and further suggested that its predation was good for coral diversity, their views were at least as extreme as those of Chesher. Perhaps we might grant Grigg some degree of poetic license in his description of "true Cassandras." However, if we are to retrospectively select "correct" elements from among theories, then we must symmetrically apply the same criterion when assessing the "false Cassandras." In hindsight, Chesher was considered to have been right in at least two fundamental respects: there was a real increase in *Acanthaster* populations, and anthropogenic factors may well be involved in the observed outbreaks. Moreover, some coral-reef scientists still believe that deterioration in coral-reef communities due to *Acanthaster* predation and other anthropogenic effects may be irreversible.

One cannot explain the long controversy in terms of a failure to employ scientific objectivity when choosing the best theory and best policy, because values, technical issues, and broader ecological theory underscore all aspects of this controversy. Facts, theories, values, and politics were so entangled in the controversy that it was often as difficult for us to separate them as it was for scientists to separate anthropogenic from natural change. To invoke "the best available evidence" as the arbitrator in this dispute is to sidestep the very issues we needed to address: how knowledge is forged in environmental science, how evidence is obtained, how it is assessed, how conclusions are reached (if indeed they are) and action taken.

The crown-of-thorns controversy is not a story about false Cassandras. It is one that belongs properly to the history of coral-reef science, to ecology, and environmental science more generally. There are many parallels in addition to coral bleaching and global warming. Sudden surprises and unexpected behaviors have challenged traditional myths of causation and environmental management.[23] Outbreaks of the spruce budworm have increased in frequency and magnitude in eastern and central Canada and the United States. During the 1960s, the outbreaks began in Newfoundland where the insect was thought to be historically rare. More recently, they have become more frequent in stands of young trees. Is ecological knowledge simply deficient or are these outbreaks really new and due to human activities: creating monocultures, managing forest fires, and using pesticides? The underlying causes of the behavior of spruce budworm outbreaks and long-term effects of management policies continue to divide ecologists as do the causes of the collapse of the sardine fishery in California in the 1940s, the herring fishery in the North Sea in the 1980s, or the cod fishery in the north Atlantic in the 1990s.[24]

In many cases, policy decisions to intervene to maintain desired species have hinged on the question of whether environmental perturbations are natural or human caused. In the crown-of-thorns controversy many activists maintained, with Chesher and Endean, that the outbreaks should be controlled whether they were natural or not. They drew an analogy with fire. Whether caused by lightning or match, no one would hesitate to put out a forest fire. Nonetheless, the opposing laissez-faire attitude was maintained on the grounds that if such disturbances were natural they may be important for a balanced ecological community and therefore should not be controlled. Strikingly, this same argument persisted even in terrestrial park management —and in the analogous case of forest fires.

Nothing provides a more vivid case of the "naturalist fallacy," that if it is natural it is good, than the fires of 1988 that burned over the Greater Yellowstone Area, the site of the first national U.S. park (1872) and its earliest national forest (1891). They are recognized today as among the most significant events in the history of national parks.[25] The fires, the largest ever recorded for that area, were allowed to spread and burn, in that sweltering summer, on the grounds that they were part of the natural processes occurring in wilderness landscapes and that both needed to be preserved as "nature intended," untouched by humans. It was also assumed, based on past experience with fire management and historical data, that large fires would be confined to old-growth pine or aging spruce–fir forests.[26] Yet, 45% of Yellowstone National Park burned.[27] The Yellowstone fires were also notorious for the intensity and scale of public and media attention given to them, the great costs of their attempted suppression, and for the test they provided for the management philosophies and policies of parks and wilderness areas.

Among the many lessons about the unpredictability of nature and the importance and difficulty of distinguishing between anthropogenic and natural causation, ecologists learned enough to know that management strategies had to be based on consequences of such perturbations. Fires that might be considered "natural" by various criteria were not necessarily acceptable, and "natural landscapes" were not "predestined" to achieve some particular ecological structure or configuration if one simply removed human influences. One had to live and manage in a dynamic symbiosis with a changing Earth.

Values could not be determined by nature. The question for the next generation of park managers was not whether manipulation is desirable, but what kind of manipulation is desirable, and how to achieve it.[28] Many ecologists recognize that most of the world's ecosystems are affected to some degree by human activities. But, as some put it bluntly, in no place can they claim to predict with certainty either the ecological effects of the activities or the efficacy of most measures aimed at regulating or enhancing desired species.[29]

The outbreaks of the crown-of-thorns starfish were central to the co-evolution of coral-reef ecology and environmentalism. The emergence of environmental awareness during the 1960s and 1970s can be clearly viewed through the struggles between and among scientists, conservationists, and politicians as they attempted to ignore, accept, define, and finally begin to understand this new phenomenon. The crown-of-thorns controversy is about coral-reef scientists struggling to come to grips with new technology, encountering new phenomena, developing theories, establishing techniques, and building institutionally. It is about public participation in environmental controversies and scientists' participation in public controversies. The crown-of-thorns controversy is indeed a microcosm for the playing out of global environmental controversies.

Despite the appeals to objectivity and "ecorealism," ecological reality, like all of science, actually leaves much to the imagination. When encountering a startling new observation, the question is often raised whether it is an artifact of new techniques employed or a new phenomenon. Coming at a time when underwater observations with scuba diving were becoming prominent, the claim that the outbreaks themselves were new and therefore "unnatural" could easily be challenged. Those who believed, in the 1960s and early 1970s, that the starfish predation represented a serious environmental threat organized leading scientists to witness the phenomenon and solicited testimonies from amateur divers throughout the Indo-Pacific. They supplied photographs and films of devastation, with expert testimonies about the age of the coral destroyed. They itemized possible anthropogenic factors and detailed the potential risks to coral reefs, fisheries, and tourism.

Yet these resources were not strong enough to resist the claim that the outbreaks were natural and the problem self-correcting. When predation data and starfish numbers were largely anecdotal, when appropriate survey techniques had not been standardized, and when there was no consensus on what constituted "normal" populations, even the claim that there really were any crown-of-thorns population increases in the Indo-Pacific could be doubted. Indeed, in the absence of monitoring networks, proving the existence of the plagues was often as difficult as understanding their causes and effects. And just as in a case of an outbreak of infectious disease, governments in tropical areas were often sluggish in exchanging information that could harm the tourist industry. Theory choice and choice of action depended on a number of considerations.

All theories about the scope and significance of the starfish plagues were propped up by social forces inside and outside the scientific community. Heavily supporting the side of "man-induced plague," newspapers and magazines kept the debate in public view. Journalists demonized *Acanthaster* as a monster preying on healthy life-giving and beautiful coral—a plague, the curse for man's reckless exploitation of the planet. The crown-of-thorns was invested with politico-economic and moral meaning. As represented in magazines and newspapers, the plagues were a measure of ever-increasing industrial and technological production, unbalanced by knowledge of our own social world and that of other species. Indeed, the crown-of-thorns was an icon representing all the fears associated with our careless exploitation of the Earth.

Like microbe hunters attacking outbreaks of deadly infectious diseases, scientists of the late 1960s and early 1970s helped to organize and equip *Acanthaster* hunters with chemical-injecting guns and spears. They told of their encounters with the enemy while encouraging others to engage in the new battle of the Pacific. To understand their actions we have had to consider their values and politics. As tropical coral reefs tend to be located in poorer nations, they needed strong advocates. Even those less confident that the plagues were caused by human activity demanded action. Some potential environmental threats, they argued, required immediate action; they could not wait for the kind of rigorous tests required of laboratory science. Many adopted the precautionary principle that control measures could not hurt even if the outbreaks were natural. On the other hand, if they were to do nothing and the outbreaks did prove harmful, their inaction would be unconscionable. Little was certain about whether the reefs would recover, or how long it would take. And Pacific islanders preferred to protect their reefs.

Calls for local control programs were intermingled with calls for research to reveal the cause and perhaps more effective control measures. *Acanthaster* pointed to an underdeveloped area of scientific research of intense public interest. It was used to explore environmental and social problems—the over-use of pesticides, pollution, overfishing, and impacts of tourism—and guide solutions to them. And it was on the basis of environmental degradation that many coral-reef research institutions, reserves, and marine parks were established. To look askance at the major publicists in the crown-of-thorns controversy (and in the coral-bleaching controversy) is to scorn the very researchers who attracted funds for such problems in the first place. Activists called for research and control measures simultaneously, while their critics in Australia and in the United States denied the need for special funds for crown-of-thorns research or for control programs.

In the United States mainland, publicity for *Acanthaster* came and went by 1972, as one journalist had predicted: "Pests like celebrities tend to emerge into the limelight and then disappear from news." But in Australia, public pressure remained, attached to larger issues about the conservation and commercial and industrial exploitation of the Great Barrier Reef. Lobbying in newspapers and on television, emphasizing the need for control measures, resulted in repeated committees of inquiry that only confirmed the extent of disagreement and inflamed the controversy instead of resolving it. Maintaining that the outbreaks were natural, the crown-of-thorns committees in Australia were more concerned with controlling public outcries than stopping the starfish outbreaks.

The rigid dichotomy between anthropogenic and natural cause was not determined by logical necessity. It was reinforced by the politico-economic nature of the controversy over the need for eradication programs. While governments and many tourist operators were reluctant to acknowledge the problem and make it known for fear of its negative impact on their industry, other tourist operators and conservationists wanted the problem recognized, tackled, and solved by (government-sponsored) control programs. This was not an issue that could be easily controlled by science. It was not simply a matter of correcting exaggerations (or denials) in regard to the scope and significance of the phenomenon. A second wave of public outcries in Australia occurred during the 1980s despite such orchestrated efforts by the Great Barrier Reef Marine Park Authority to manage the public controversy.

The crown-of-thorns problem was as important for understanding fundamental aspects of reef ecology as for emerging environmental policy. The conditions for such ecological studies were difficult and complex. Before the crown-of-thorns could be placed in a thorough ecological context, marine ecologists had to shed their "fuzzy" attribution and establish themselves solidly within institutionalized science and advisory bodies of government. New marine laboratories had to be constructed in tropical locations for full-time coral-reef researchers. Within ecologists' circles, the expressions "plagues" and "infestations," loaded with prejudicial implications, were emptied from the "*Acanthaster* phenomenon" and replaced by the term "outbreaks." At the same time, the rhetorical dichotomy between natural and anthropogenic cause began to erode.

Scientists converged on the crown-of-thorns from diverse specialties, often with divergent approaches, theories, and techniques. Geologists, natural historians, community ecologists, and modellers co-existed in varying degrees of cooperation and conflict. As in all science, investigators contended over what questions are important, what answers are acceptable, what techniques are most useful and what phenomena are most interesting. As we followed scientists into their laboratories, out on expeditions, into meetings, and in debates in technical journals, we saw how they scrutinized each others' methods, observations, and conclusions, often combining and refashioning them in complex ways.

The ecology of global outbreaks revealed itself to be more complex than fuzzy. Early in the controversy it was agreed that any satisfactory theory had to account for all the outbreaks occurring almost simultaneously throughout the Indo-Pacific. One had to consider a plethora of variables: geographical and hydrological conditions at the time; the behavior of the organism at various stages of its life history and in diverse conditions; predator–prey relations; and human activities in coastal areas. Even then, there was no agreement as to what constituted the best facts or approach. Geologists based their evidence on historical interpretations of sediment records; some naturalists and ecologists focused on the attributes of the starfish and its relations with the physical environment; others emphasized predator–prey relations, while still others sought an understanding in long-term scales of hundreds of years. Within this competitive matrix, data were deployed and analyzed with differing degrees of scepticism and acceptance. There was no key approach strong enough to close this controversy.

Historical evidence of past outbreaks, whether by the qualitative methods of the social historian or the quantitative studies of geologists, was questionable. Anecdotal statements from past scientific expeditions, assertions from local fishermen, or cultural linguistics were simply that. Alternative interpretations could be given to the geological evidence. Even then, critics argued, evidence for past outbreaks foretold little with certainty about the cause of present outbreaks.

Laboratory-based knowledge on the biology of *Acanthaster* was also not sufficient for constructing a theory about the cause of the outbreaks. Facts acquired from the controlled conditions of the laboratory had to be tested in the field. There were difficulties of getting the time and location of events in the field to match. Data regarding optimal salinity levels and temperature for larval development were important. But such conditions were too common. Excessive nutrient runoff caused by typhoons following periodic droughts was possibly a key predictive factor. Birkeland's field evidence from Micronesia was matched by Lucas's laboratory experiments in Townsville. But critics argued that it was difficult to obtain evidence for matching weather conditions in all outbreak areas, especially the Great Barrier Reef. And why were other species of starfish not affected by nutrient runoff?

Overfishing could not be overlooked, but field studies of population dynamics lagged behind laboratory studies. The giant triton was shown to be a predator of adult *Acanthaster*, but its predation seemed to be inadequate and its normal numbers unknown. Other starfish predators were known, but their normal numbers in outbreak regions were unknown—at least until the late 1980s.

One also had to consider fundamental conceptual issues about complex systems that had polarized debates about the cause and potential effects of the outbreaks. The belief that complex tropical communities were more stable than less complex ones was virtually non-problematic—that is, unquestioned among ecologists during the first half of the century. It remained uncontested by snapshot images of coral reefs carried away by expeditions. But it was disputed during the 1970s and 1980s.

Mathematical ecologists and theoretical community ecologists turned to study longer-term and wider-scale multispecies ecology and challenged the accepted relation between complexity and stability. Non-equilibrium models were constructed to account for species diversity in coral reefs and tropical rain forests in terms of stochastic (random) changes. The crown-of-thorns outbreaks fit such models—inasmuch as the plagues could be considered natural. That view was accompanied and supported by other institutional changes. As permanent tropical laboratories emerged, so too did observations of extinctions, mass mortalities, and other kinds of outbreaks in those regions. One could not simply assume they were all novel events due to recent human activities. To discern the causes and effects of those changes on ecological communities, ecologists had to contemplate processes that lasted longer than their own research careers and over areas far too large for their conventional experiments.

The relation between species diversity and stability required further empirical study, as did *Acanthaster* outbreaks. Indeed, the recognition that such outbreaks *may* have occurred naturally in ecologically complex systems did not mean that the observed outbreaks *were* natural. But it did mean that any account precluding the possibility of outbreaks in the past was severely weakened. Both perspectives could be right: the outbreaks were not new, nor were they pristine processes.

Disequilibrium theory brought with it the notion that some "intermediate" disturbances may be good for biodiversity. But, by 1980, most coral-reef ecologists denied that the observed crown-of-thorns outbreaks had such an effect on coral diversity. They also excluded the outbreaks from being simply part of long-term natural cycles. When the first wave of outbreaks dissipated by the mid-1970s, it seemed to indicate that they were the passing periodic oscillations that some had long claimed them to be. But their sudden reappearance over the next decade was at odds with that view. Developments along coastal areas causing increased runoff could have increased the frequency and magnitude of outbreaks. That overfishing may also be involved was shown by comparative studies of fish predator densities in areas where starfish outbreaks were common and in areas where they were unknown to occur.

Perhaps many of the factors entertained in all the main theories—increased larval nutrition, adult aggregation, predator–prey relations, including overfishing—acted together to create outbreak conditions.

During the 1990s, coral-reef scientists globally became ever more concerned with coral reef damage, degradation, depletion, and destruction ultimately due to rapid population growth in tropical developing countries migration to coastal areas, together with technological developments for reef exploitation. However, as phenomena such as coral bleaching, the mass die-off of *Diadema* in the Caribbean, and crown-of-thorns outbreaks indicated, measuring such effects and managing conservation had to be done against the background of large-scale ecological processes. How to distinguish between human-induced perturbations and hypothetical long-term natural oscillations continues as the crucial question of the hour as coral-reef scientists grapple with the problem of how to approach global environmental changes on the temporal and spatial scales at which nature operates.

Currently, crown-of-thorn outbreaks are chronic on the reefs at Fiji, Guam, Palau, and the Ryukyu Islands. In April 1996, Birkeland visited Palau and resurveyed the areas of outbreaks of 1977–78. "These areas have not only failed to recover after nearly 20 years, but they have deteriorated further. I think it is overfishing of herbivorous fishes so the coral recruits can't get a start. The areas are all covered with algae. This is in great contrast to the beautiful healthy reefs which COTS did not infest in the 1970s." [30] *Acanthaster* has also been blamed for die-off in corals in large numbers in Sri Lanka, the Maldives, Mauritius, Malaysia, and Indonesia.[31] The second series of outbreaks on the Great Barrier Reef subsided by 1990. There is word now that a third series is making its way southward from the middle of the Great Barrier Reef.

NOTES

Introduction

1. Our propensity to believe such accounts, they speculate, may be due to a variety of social and psychological issues ranging from paranoid delusions, fear of our own individual deaths, to our "need for enemies." See, for example, Ben Bolche and Harold Lyons, *Apocalypse Not: Science, Economics and Environmentalism* (Washington, D.C.: The CATO Institute, 1993); Aaron Wildavsky, *But is it True? A Citizen's Guide to Environmental Health and Safety Issues* (Cambridge: Harvard University Press, 1995); Robert Jastrow, William Nierenberg, and Frederick Seitz, *Scientific Perspectives on the Greenhouse Problem* (Chicago: The Marshall Press, 1990). See also David L. Bender and Bruno Leone, eds., *The Environmental Crisis: Opposing Viewpoints* (San Diego: Greenhaven Press, 1991).

2. Andrew Nikiforuk, *The Fourth Horseman. A Short History of Plagues, Scourges and Emerging Viruses* (London: Penguin, 1992); Richard Preston, *The Hot Zone* (New York: Random House, 1994); Robin Henig, *A Dancing Matrix: Voyages During the Viral Frontier* (New York: Knopf, 1993); Laurie Garrett, *The Coming Plague: Newly Emerging Diseases in a World out of Balance* (New York: Farrar, Straus and Giroux, 1994); Rodney Barker, *And the Waters Turned to Blood: The Ultimate Biological Threat* (New York: Simon and Schuster, 1997).

Chapter 1

1. John H. Barnes, "The Crown of Thorns Starfish as a Destroyer of Coral," *Australian Natural History* 15 December (1966):257–61; 257.

2. Ibid., 258.

3. Robert Raymond, *Starfish Wars: Coral Death and the Crown-of-Thorns* (Melbourne: Macmillan, 1986), 18–20.

4. Ibid., 29–30.

5. Barnes, "The Crown of Thorns," 259.

6. Ibid., 260.

7. Ibid., 258.

8. Ibid., 257.

9. Ibid., 259.

10. Ibid., 261.

11. To add to this ecological upheaval, Barnes suggested that the starfish might possibly be the root cause of some cases of ciguatera poisoning that were increasing in nearby areas. Ciguatera-type neurotoxins had been found to be increasingly present in the flesh of some species of fish. It was well known that normally edible fish can, on occasion, suddenly become toxic.

This appeared to have been happening in Townsville, 120 miles north of Cairns. Barnes thought that some of the algae, which gathers on the dead coral after a starfish onslaught, might be toxic. If so, the toxin would be transmitted via fish that feed on the algae to humans, causing the food poisoning. Ibid., 260.

12. For an overview of the early development of coral-reef research, see C. M. Yonge, "The Biology of Coral Reefs," *Advances in Marine Biology* 1 (1963):209–59.

13. Ibid.

14. Ibid.

15. Ibid.

16. Ibid., 209.

17. Barnes, "The Crown of Thorns Starfish," 261.

18. J. H. Barnes and R. Endean, "A Dangerous Starfish—*Acanthaster planci (Linné),*" *Medical Journal of Australia* 1 (1964):592–93.

19. Robert Endean, "Report on Investigations made into Aspects of the Current *Acanthaster planci* (crown-of-thorns starfish) Infestations of Certain Reefs of the Great Barrier Reef," Fisheries Branch, Queensland Department of Primary Industries, Brisbane, Australia, 20 April 1969. This was followed by a more technical report by R. G. Pearson and R. Endean, "A Preliminary Study of the Coral Predator *Acanthaster planci* on the Great Barrier Reef," *Fisheries Notes,* Department of Harbours and Marines, vol. 3, no. 1. Queensland Government, December 1969, 27–55.

20. Richard Kenchington, "The Crown-of-thorns Crisis in Australia: A Retrospective Analysis," *Environmental Conservation* 5 (1978):11–19; 12.

21. Endean, "Report on Investigations," 12.

22. Ibid.

23. Ibid., 10.

24. Ibid., 18.

25. Ibid., 16.

26. Ibid., 16. Commercial shell collecting had also been carried out at localities outside the Great Barrier Reef where the crown-of-thorns had only recently become plentiful: Rabaul, Manus island, Fiji, and Western Samoa. See also anon., "Ban the Triton Shell Collectors," *Pacific Islands Monthly,* February (1970):37–39.

27. Ibid., 17.

28. Ibid.

29. Ibid., 21.

30. Ibid.

31. V. J. Loosanoff and J. B. Engle, "Chemical Control of Starfish," *Science* 88 (1938):107–109; idem, "Use of Lime in Controlling Starfish," *U.S. Fish and Wildlife Service Research Department* 2 (1942):1–29.

32. Endean, "Report on Investigations," 22. Endean (p. 34) summarized his recommendations to the Queensland government with an estimated cost

of about $250,000 for the first two years:

1. A program aimed initially at containing the *A. planci* plague be instituted as soon as possible, under the direction of a senior scientist.
2. Collectors be employed to remove starfish first from reefs peripheral to the area of *A. planci* infestations and then from the reefs within the main area of infestation.
3. Local organizations along the coast be asked to provide volunteers to help collect starfish.
4. A research team continue to study the biology of *A. planci*, its main predator, the giant triton, with live tritons being imported for field study purposes and for the re-stocking of the reefs.
5. The ban on the taking of tritons on the Great Barrier Reef be rigidly enforced, and consideration given to restricting the activities of both amateur and commercial shell collectors on the reefs.

33. Anon., "Giant Starfish Attacks Great Barrier Reef," *Science Journal* 2 (1966):9–10; D. E Williamson, "The Coral Predator," *Skin Diver* 17 (1968): 26–27.

34. Jon Weber, "Disaster at Green Island—Other Pacific Islands May Share Its Fate," *Earth and Mineral Sciences* 38 (1969):37–41, 37.

35. The first reference Weber could find to the starfish was made by Rumphius in 1705, who referred to it as *Stella marina quindecium radiorum* (*Latin*, "fifteen rayed sea star"). See P. de Loriol, "Echinodermes de La Baie d'Ambouie," *Rev. suis. zool. et Ann. d. Mus. d'Hist. nat. Geneve* 1 (1893): 359–426. Others traced its name back further to an animal called *Stella pentekaidekaktis* (*Latin–Greek*, "star comb of fifteen") by Columna in 1616. See J. Muller and F. H. Troschel, *System der Asteriden* 1842. See also John Ellis, *The Natural History of Many Curious and Uncommon Zoophytes Collected from various Parts of the Globe*. Arranged by Daniel Solander (London: Benjamin White and Son, 1786). The name *Asterias planci* was bestowed upon it by Linnaeus in 1748 for a specimen described by Plancus in 1743.

36. See F. J. Madsen, "A Note on the Sea-Star Genus Acanthaster," *Vidensk. Medd. Dansk naturh. Foren.*, 117 (1955):179–92. Ibid.

37. H. L. Clark, "The Echinoderm Fauna of Torres Strait: Its Composition and Its Origin," *Carnegie Institution of Washington Publications* 214 (1921):1–224.

38. W. K. Fisher, "Starfishes of the Philippine Seas and Adjacent Waters," *United States Natural History Museum Bulletin* 3 (1919):32–35.

39. A. A. Livingstone, "Great Barrier Reef Expedition 1928–29," *Scientific Reports 4, no.8, Asteroidea* (1932).

40. W. K. Fisher, "Marine Zoology of Tropical Central Pacific," *B. P. Bishop Museum Bulletin 27. Tanger Expedition Publication No. 1* (1925): 81–82.

41. Weber, "Disaster at Green Island," 38.

42. Ibid.

43. Ibid., 39.

44. Ibid., 40.

45. Ibid., 38.

Chapter 2

1. Richard H. Chesher, "Divers Wage War on the Killer Star," *Skin Diver*, 18 March (1969): 34–35, 84–85; 34.

2. Ibid., 84–85.

3. Richard Chesher,"Destruction of Pacific Corals by the Sea Star *Acanthaster planci*," *Science* 165 (1969):80–83; 283.

4. Lu Eldredge, interview, Bernice P. Bishop Museum, Honolulu, Hawaii, April 18, 1995.

5. Richard Chesher, "*Acanthaster*: Killer of the Reef," *Oceans* 3 (1970): 11–17; 14.

6. Wesley R. Coe, "Fluctuations in Populations of Littoral Marine Invertebrates," *Journal of Marine Research* 15 (1956–57):212–32; 219.

7. Chesher, "*Acanthaster*: Killer of the Reef," 14.

8. E. B. Worthington, ed., *The Evolution of the IBP* (Cambridge: Cambridge University Press, 1975), 6.

9. As its founders later saw it, the IBP paralleled the "Environmental Revolution." At the same time it helped transform ecology from a descriptive science to one with a dynamic ecosystems approach, examining energy processing and productivity. Their kind of thinking was the same that had stimulated such world activities as the Biosphere Conference at Paris in 1968, which led to the Man and Biosphere Program, and the UN Conference on the Human Environment at Stockholm in 1973 which led to the creation of The United Nations Environmental Program (UNEP). Ibid., xvii–xviii.

10. Ibid., 94.

11. C. M. Yonge, "Living Corals," *Proceedings of the Royal Society* B 169 (1968):329–44.

12. T. F. Goreau, "On the Predation of Coral by the Spiny Starfish *Acanthaster planci* (L.) in the Southern Red Sea," *Israel South Red Sea Expedition, 1962, Sea. Fish. Res. Stat. Haifa, Bulletin No. 35* (1–4), 23–26.

13. Ibid., 26.

14. Chesher, "*Acanthaster*: Killer of the Reef," 14.

15. Ibid.

16. Chesher, "Divers Wage War," 84.

17. Chesher, "Destruction of Pacific Corals," 280.

18. Chesher, "*Acanthaster*: Killer of the Reef," 14.

19. Ibid.

20. Ibid.

21. See Thomas F. Goreau, Nora I. Goreau and Thomas J. Goreau, "Corals and Coral Reefs," *Scientific American* 241 (1979):124–36; 124.

22. Chesher, "*Acanthaster:* Killer of the Reef," 12.

23. Ibid.

24. Ibid., 15.

25. Chesher, "Destruction of Pacific Corals."

26. Ibid., 280.

27. Ibid., 281–82.

28. Ibid., 282.

29. Ibid.

30. Ibid., 283.

31. Richard Chesher to the author, January 19, 1998.

32. The staff of Westinghouse research laboratories numbered over 1700, approximately 40% scientists and engineers. The Westinghouse library maintained a collection of more than 60,000 volumes of technical books and periodicals, in fields of physics, chemistry, electronics, metallurgy, mechanics, mathematics, and related subjects; and subscribed to 1000 scientific and technical journals. Anon., "Assessment and Control of Coral-Destroying Starfish *Acanthaster planci* in the Pacific Ocean," Proposal to U.S. Department of the Interior Washington, D.C., submitted by Westinghouse Ocean Research Laboratory of the Westinghouse Laboratories, 20. Copy obtained from the papers of David Pawson, National Museum of Natural History, Smithsonian Institution, Washington, D.C.

33. For brief accounts of the origins of the Scripps Institution of Oceanography, see E. N. Shor, "How the Scripps Institution Came to San Diego," *Journal of San Diego History* 27 (1981):161–73; E. L. Mills, "'Useful in Many Capacities': An Early Career in American Physical Oceanography," *Historical Studies in the Physical and Biological Sciences* 31 (1990): 265–311. See also idem, *Biological Oceanography: An Early History, 1870– 1960* (Ithaca: Cornell University Press, 1989); Idem, "The History of Science and Oceanography after Twenty Years," *Earth Science History* 12 (1993):5–18.

34. For example, WORL's studies of the physical and biological implications of the heated hypersaline effluent discharged into the ocean by a power and desalting plant were partially supported by the Federal Water Pollution Control Administration. Its development of a computerized technique for modeling resonant forced wave oscillations in semi-closed basins was supported by the Coastal Engineering Research Center of the U.S. Army Corps of Engineers.

35. Ibid., 22.

36. Ibid., 1.

37. Ibid., 12.

38. Richard H. Chesher, "*Acanthaster planci*: Impact on Pacific Coral Reefs," Final Report to U.S. Department of the Interior, Washington D.C.; Research Laboratories, Westinghouse Electric Corporation, Pittsburgh, Pennsylvania, October 15, 1969, PB187631. Reproduced by the Clearinghouse for Federal Scientific and Technical Information, Springfield, Va. 22151; ii–iii. Copy obtained from the papers of David Pawson, National Museum of Natural History.

39. Many leading ecologists in the United States can trace their intellectual ancestry directly or indirectly to the Hutchinson school. See Yvette H. Edmondson, "Some Comments on the Hutchinson Legend," *Limnology and Oceanography* 16 (1971):157–89. Hutchinson's well-ornamented phylogenetic tree of intellectual descendants is illustrated on page 162.

40. As Yonge put it in 1963, his work had "completely altered our understanding of both aspects of [coral] growth." See C. M. Yonge, "The Biology of Coral Reefs," 209–60; 247.

41. C. M. Yonge, "The Biology of Reef-Building Corals," *Scientific Reports of the Great Barrier Reef Expedition. 1928–1929*, 1 (1940):353–91; idem, "Experimental Analysis of the Association between Invertebrates and Unicellular Algae," *Biological Reviews* 19 (1944):68–80; idem, "Symbiosis," *Memoranda of the Geological Society of America* 67 (1957):429–42.

42. T. F. Goreau, "The Physiology of Skeleton Formation in Corals. I. A Method for Measuring the Rate of Calcium Deposition by Corals Under Different Conditions," *Biological Bulletin* 116 (1959):59–75; idem, "Problems of Growth and Calcium Deposition in Reef Corals," *Endeavour* 20 (1961):32–39; T. F. Goreau, and N. I. Goreau, "Distribution of Labelled Carbon in Reef-Building Corals with and without Zooxanthallae," *Science* 131 (1960):668–69.

43. Jeremy Jackson, interview, Panamá, June 26, 1996.

44. D. J. Barnes, "Coral Skeleton: an Explanation of their Growth and Structure," *Science* 170 (1970):1305–08. See also D. J. Barnes and J. M. Lough, "Systematic Variations in the Depth of Skeleton Occupied by Coral Tissue in Massive Colonies of Porites from the Great Barrier Reef," *Journal of Experimental Marine Biology and Ecology* 159 (1992):113–28; D. J. Barnes and J. M. Lough, "On the Nature and Cause of Density Banding in Massive Coral Skeletons," *Journal of Experimental Marine Biology and Ecology* 167 (1993):91–108.

45. Jackson, interview.

46. Jeremy Woodley, interview, University of the West Indies, Kingston, Jamaica, January 29, 1997.

47. Ibid.

48. Jackson, interview.

49. Jeremy Woodley, note to the author, August 2, 1996.

50. Woodley, interview.

51. Chesher, "*Acanthaster planci*: Impact on Pacific Coral Reefs," 2.

52. Ibid., 15.

53. Jackson, interview.

54. Ibid. Some members of the Truk team noted that the starfish did not appear to eat all corals equally. They seemed to prefer *Acroporas* and other fast-growing coral. See, D. J. Barnes, R. W. Bauer, and M. R. Jordan, "Locomotory Response of *Acanthaster planci* to Various Species of Coral," *Nature* 228 (1970):342. R. W. Bauer, M. R. Jordan, and D. J. Barnes, "Trig-

gering of the Stomach Eversion Reflex of *Acanthaster planci* by Coral Extracts," *Nature* 228 (1970):344 .

55. Chesher, "*Acanthaster planci*: Impact on Pacific Coral Reefs," 33.

56. Ibid.

57. Ibid., 16.

58. Ibid., 17.

59. Ibid., 3.

60. Ibid., 33.

61. Ibid., 95–99.

62. Ibid., 95.

63. Ibid., 97.

64. Ibid., 99.

65. Ibid., 99.

66. Chesher thought it "a conservative assumption to believe that the population explosions, in one way or another, correlate to human activities." Ibid., 118–19.

67. Ibid., 101.

68. Ibid., 107.

69. Ibid., 110.

70. Ibid., 110–11.

71. Ibid., 112.

72. Ibid., 114.

73. Ibid., 116.

74. Ibid., 120.

75. Ibid., vi.

76. Ibid., 121.

77. Ibid., 126.

78. Ibid., 127.

79. Ibid., 122.

80. David Alexander, "Guarding a Watery Paradise," *International Wildlife*, 18 no. 3 (1988):4–11; 6.

81. Sue Wells, "A Future for Coral Reefs," *New Scientist* 30 October (1986), 46–50; 50.

82. Chesher, "*Acanthaster planci*: Impact on Pacific Coral Reefs," 123.

83. Ibid., 123.

84. Ibid., 124.

85. Ibid., 126.

86. Ibid., 128.

87. Ibid., 130.

88. Ibid., 130.

89. Ibid., 131.

90. Anon., "Summary of 'Crown of Thorns' Workshop Discussions, October 9–10, 1969, University of California, San Diego, Scripps Institution of Oceanography." Obtained from library of David Pawson, National Museum

of Natural History, Smithsonian Institution, Washington, D.C., 4.

91. Ibid., 5.

92. Ibid., 6.

93. Anon., "Crown of Thorns Investigation," *BioScience* 20 (1970):113.

94. The Committee on Merchant Marine and Fisheries, "Crown of Thorns Starfish," House of Representatives, August 11, 1970, Report No. 91–1406, 3. Copy obtained from the papers of David Pawson, Museum of Natural History, Washington, D.C.

95. Ibid.

96. "The bill would authorize to be appropriated not in excess of $4.5 million—from date of enactment of the bill to June 30, 1975—for research on and ways to control the Crown of Thorns Starfish (*Acanthaster planci*) which have completely destroyed miles of coral reefs." Ibid., 1.

Chapter 3

1. H. G. Wells, *The War of the Worlds* (London: Heinemann, 1898).

2. D. C. Smith, *H. G. Wells: Desperately Mortal* (New Haven: Yale University Press, 1986), 64–67.

3. Ibid., 65.

4. Charles E. Elton, *The Ecology of Invasions by Animals and Plants* (London: Methuen and Co. Ltd., 1957).

5. Barry Commoner, *Science and Survival* (London: Gallancz, 1967); Paul R. Ehrlich, *The Population Bomb* (Rivercity, Mass.: Rivercity Press, 1975).

6. Rachel Carson, *Silent Spring* (New York: Houghton Mifflin Co., 1962), 252.

7. Richard Chesher, "Divers Wage War on the Killer Star," *Skin Diver* 18 (1969):34–35, 84–85; 34.

8. Ibid., 85.

9. Richard H. Chesher, "*Acanthaster*: Killer of the Reef," *Oceans* 3 (1970): 11–17.

10. Ibid., 17.

11. L. G. Eldredge, *Acanthaster Newsletter* No 2., July 1970, 5pp. Copy obtained from the library of David Pawson, National Museum of Natural History, Smithsonian Institution, Washington, D.C.

12. Ibid., 2.

13. W. Wickler and U. Seibt, "Das Verhalten von Hymenocera picta Dana, einer Seesterne fressenden Garnele" (Decadapoda, Natania, Gnathophyllidae). *Zeitschrift für Tierpsychologie* 27 (1970):352–68.

14. Anon., "The Starfish Eaters," *Time*, 25 May (1970):73.

15. Ibid.

16. Eldredge, *Acanthaster Newsletter*, 3.

17. Anon., "Plague in the Sea," *Time*, 12 September (1969):57.

18. Milton MacDonald, "The Ambush at Puntan Muchot," *Micronesian*

Reporter, Second Quarter (1970): 1, 9–13.

19. Anon., "The Starfish Plague," *The Illustrated London News*, 21 November (1970):19–21.

20. G. G. T. Harrison, "The Crown-of-Thorns Starfish," *The Fisherman* (Official Journal of State Fisheries of New South Wales) December (1969):9–10.

21. Douglas E. Williamson, "Does the Great Barrier Reef Face Destruction by the Deadly Crown of Thorns?" *Skin Diver*, March (1968):26–27; 26.

22. Ibid., 27.

23. Stuart Auerbach, "Coral-Eating Starfish Peril Pacific Isles," *Washington Post*, June 12, 1969, A15.

24. Anon., "Battle of the Coral Sea," *Newsweek*, 14 July (1969):53.

25. Ibid.

26. David S. Landes, *The Unbound Prometheus: Technological Change and Industrial Development in Western Europe from 1750 to the Present* (Cambridge: Cambridge University Press, 1969), 554–55.

27. See Isaac Asimov, *The Intelligent Man's Guide to Science, Volume II, The Biological Sciences* (New York: Basic Books, 1960).

28. D. J. de Solla Price, "Review of Asimov's Text," *Science* 132 (1960):1830.

29. E. B. Worthington, ed., *The Evolution of the IBP* (Cambridge: Cambridge University Press, 1975), 7.

30. Ibid., 9.

31. Today, the cheering for the intellectual, financial, and medical merits of the human genome project is met by jeering cries of hype by critics and the follies of centralized top-down big science that threatens the diversity of smaller research efforts. See, for example, Eliot Marshall, "Less Hype, More Biology Needed for Gene Therapy," *Science* 270 (1995):1751; Joshua Lederberg, "What the Double Helix (1953) Has Meant for Basic Biomedical Science," *Journal of the American Medical Association* 269 (1993):1981–1985. See also, Daniel Kevles and Leroy Hood, eds., *The Code of Codes* (Cambridge: Harvard University Press, 1992). Some argue that the rise of such gene-eyed views of humans simply serve a conservative social agenda, reflecting society's eagerness to blame ill health and misfortune on individuals rather than on social and environmental conditions. They deflate the grandiose promises of therapeutic benefits supposed to emerge from the human genome initiative, and instead point to the real threats to privacy and liberties already resulting from the unregulated increase in genetic information and predictions. See, for example, Ruth Hubbard and Elijah Wald, *Exploding the Gene Myth* (Boston: Beacon Press, 1993). At the same time, ecologist Stuart Pimm argues that "understanding large scale ecological processes is far more of an intellectual challenge than is the stupefying tedious sequencing of the human genome. The problems are also more important." See Stuart Pimm, *The Balance of Nature? Ecological Issues in the Conservation of Species and*

Communities (Chicago: The University of Chicago Press, 1991), xi.

32. Robert Trumbull, "Scientists Say Coral-Eating Starfish Peril Pacific Islands," *New York Times*, July 21, 1969, C 35. The same article appeared in *Miami News*, "'Starfish Invasions' poses Threat," July 24, 1969, 8-C.

33. Ibid.

34. "Task Force Studies St. Louis Airshed," *Centre for the Biology of Natural Systems* 3(1970):3–4.

35. Anon., "Starfish Infestation Affects Pacific Coral Reefs," *Centre for the Biology of Natural Systems* 3 (1970):4–8,12.

36. Anon., "South Pacific Nightmare," *The Economist*, 22 November (1969):73.

37. Ibid.

38. Ibid.

39. Ibid.

40. Anon., "Giant Starfish Devour Pacific Reefs," *Miami Herald*, July 17, 1969, 10-F.

41. Stuart Auerbach, "Guam Losing Fight to Starfish," *Washington Post*, March 19, 1970.

42. Ibid.

43. Leonard Bickel, "Reef Victim of a Blunder," *Miami News*, September 18, 1969, 12-B.

44. Stuart Auerbach, "Starfish Eat Reefs Because Man Has Upset Nature, Study Says," *Washington Post*, December 10, 1969, A-13.

45. Anon., "Giant Starfish Devour Pacific Reefs," 10-F.

46. Anon., "South Pacific Nightmare."

47. Bernard Dixon, "Doomsday for Coral," *New Scientist* 44 (1969):226–27; 226.

48. Anon., "Research Shows Starfish Threat is Extremely Serious," *Ocean Industry*, February (1970):52–54; 52.

49. Quotations in Richard A. Kenchington, "The Crown-of-thorns Crisis in Australia: A Retrospective Analysis," *Environmental Conservation* 5 (1978):11–20, 13–14.

Chapter 4

1. Richard Chesher, "Status of *Acanthaster* Research May 1970," unpublished report, obtained from David Pawson papers, Museum of Natural History, Smithsonian Institution, Washington D.C.

2. Ibid., 1–2.

3. Ammonium hydroxide was considered superior to formalin because it was easier to handle and it is a naturally occurring compound. Ibid., 3.

4. Ibid., 4.

5. Ibid.

6. Ibid.

7. Ibid.

8. Ibid., 5.

9. Ibid.

10. Ibid.

11. Ibid.

12. Robert Raymond, *Starfish Wars. Coral Death and the Crown of Thorns* (Melbourne: MacMillan Co., 1986), 55.

13. Ibid.

14. This point has been made by many commentators: see Richard Kenchington, "The Crown-of-thorns Crisis in Australia: A Retrospective Analysis," *Environmental Conservation* 5 (1978):11–20; 18, and Raymond, *Starfish Wars*, 56.

15. Raymond, *Starfish Wars*, 56.

16. Ibid.

17. Theo Brown, with Keith Willey, *Crown of Thorns: The Death of the Great Barrier Reef?* (Sydney: Angus and Robertson, 1972). Brown's book was the first one dedicated to criticizing the Queensland Government for not taking action. See also Raymond, *Starfish Wars* and Peter James, *Requiem for the Reef. The Story of Official Distortion about the Crown-of-thorns Starfish* (Brisbane: Foundation Press, 1976).

18. Brown, *Crown of Thorns*, 14.

19. Ibid., 18.

20. Ibid., 26.

21. B. W. Halstead, "Poisonous and Venomous Marine Animals of the World," Vol 1. *Invertebrates.* U.S. Government Printing Office, L.C.-65-60000 (1965): 537–662.

22. Brown, *Crown of Thorns*, 26.

23. Ibid., 30.

24. Ibid., 31.

25. Ibid., 71.

26. Ibid., 16.

27. Roger Bradbury, interview, Panamá, July 25, 1996.

28. Ibid.

29. Ibid.

30. Ibid.

31. Ibid.

32. David Alexander, "Guarding a Watery Paradise," *International Wildlife*, 18 no. 3 (1988):4–10; 6.

33. Raymond, *Starfish Wars*, 53.

34. Alexander, "Guarding a Watery Paradise," 6.

35. Bradbury, interview.

36. Raymond, *Starfish Wars*, 53.

37. W. G. Maxwell, *Atlas of the Great Barrier Reef* (Amsterdam: Elsevier, 1968). When, in 1970, a Royal Commission was held to inquire about oil drilling, Maxwell represented oil interests. See Raymond, *Starfish Wars*, 53.

38. R. J. Walsh, M. F. Day, C. W. Emmens, D. Hill, D. F. Martyn, W. P. Rogers, D. F. Waterhouse, "*Acanthaster planci* (Crown of Thorns Starfish) and the Great Barrier Reef," *Reports of The Australian Academy of Science* Number 11 (February 1970), 3.

39. J. M. Thomson, "Marine Biology," *The Australian University* 14 (1976):114–22.

40. Vic McCristal, "Stop! The Reef Needs Time," *Walkabout* 36 (1970):26–31.

41. Walsh et al., "*Acanthaster planci* (Crown of Thorns Starfish) and the Great Barrier Reef," 8.

42. Ibid., 7.

43. Ibid., 7.

44. Ibid., 9.

45. Ibid., 12.

46. Ibid., 12.

47. Ibid., 15.

48. Ibid., 16. The Academy's report was widely publicized. See, for example, anon., "Plague Still Rages," *Nature* 226 (1970):498–99; 498.

49. Anon., "Protecting Great Barrier Reef," *Australian Fisheries*, February (1970):7.

50. James, *Requiem for the Reef*, 18. Journalists often oversimplified the conclusions of Chesher's report for the Department of the Interior, taking possibilities as facts, and exaggerated the amount of funds Americans had spent on eradication efforts. For example, Raymond, *Starfish Wars*, 55, falsely claims that the government of Guam had allotted $500,000 for control by divers using formalin injection to kill the starfish. The actual figure is $US15,000, according to Charles Birkeland and John Lucas, *Acanthaster planci: Major Management Problem of Coral Reefs* (Boca Raton: CRC Press, 1990), 8.

51. See James, *Requiem for the Reef*, 18.

52. Anon., "The Starfish Population Explosion," *Awake!*, 8 December (1970): 7–8.

53. Ibid., 8.

54. Anon., "The Starfish Could Bring 'Starvation' to the Pacific," *Pacific Islands Monthly*, April (1970):101; anon., "The Crown of Thorns Is Still Prickly," *New Scientist* 48 (1970):492.

55. Ibid.

56. James, *Requiem for the Reef*, 18.

57. Anon., "Ban the Triton Shell Collectors," *Pacific Island Monthly*, February (1970):37–39; 38.

58. Ibid.

59. McCristal, "Stop! The Reef Needs Time."

60. Ibid., 26.

61. Ibid., 31.

62. Ibid., 27.

63. Ibid., 31.

64. Ibid.

65. Ibid.

66. Ibid.

67. Anon., "Australia Warned About Environment," *Australian Fisheries* 29 (1970):12.

68. Ibid.

69. Anon., "Protecting Great Barrier Reef."

70. Raymond, *Starfish Wars*, 17, 20.

Chapter 5

1. Anon., "Committee to Study Starfish," *Australian Fisheries* 29 (1970): 12.

2. J. Walsh, C. J. Harris, J. M. Harvey, W. G. H. Maxwell, J. M. Thomson, and D. J. Tranter, "Report of the Committee Appointed by the Commonwealth and Queensland Governments on the Problem of the Crown-of-thorns Starfish (*Acanthaster planci*)" 25 March 1971 (Melbourne, CSIRO), 3.

3. Ibid., 12.

4. Richard Chesher, "Status of *Acanthaster* Research May 1970," unpublished report, 5, obtained from David Pawson papers, Museum of Natural History, Smithsonian Institution, Washington, D.C.

5. James, *Requiem for the Reef*, 57–58.

6. Anon., "Another Report on the Starfish—They're Harmless!," *Pacific Islands Monthly*, June, 1970, 41.

7. J. N. Weber and P. M. J. Woodhead, "Ecological Studies of the Coral Predator *Acanthaster planci* in the South Pacific," *Marine Biology* 6 (1970):12–17; 16.

8. Ibid., 17.

9. Ibid.

10. Theo W. Brown with Keith Willey, *Crown of Thorns: The Death of the Great Barrier Reef?* (Sydney: Angus and Robertson, 1972), 76.

11. Ibid., 76–77.

12. Ibid., 13.

13. Walsh et al., "Report of the Committee," 20.

14. "Mr. Pearson advised that he disagreed with several of Dr. Endean's interpretations of the results obtained during the investigation. . . . These objections had the effect of questioning the major conclusions drawn in the report from the limited data then available." Ibid., 49.

15. Ibid., 19.

16. Ibid., 6–7. The committee's findings were summarized in 14 concluding statements:

1. The Committee is of the opinion that the crown-of-thorns starfish, *Acanthaster planci*, does not constitute a threat to the Great Barrier Reef as a whole.

2. There is no danger of substantial erosion of the physical structure of the Great Barrier Reef and no threat to the Queensland coastline or ports. The entire living cover, or even a large portion of the coral cover of the reef, will not disappear as a result of *A. planci* activity.

3. The starfish has caused extensive damage to coral, but serious damage is limited to some reefs between the latitudes of Cairns and Townsville. It is emphasized that this damage is not uniform through this region.

4. Recolonization and regeneration of coral have occurred on all reefs that have been examined. The rate of recovery and the diversity of species are variable. Of the 39 genera of coral recorded on Green Island before the starfish infestation, 34 could still be found in June, 1970. There is no evidence that the new coral is being seriously damaged by juvenile starfish or residual adults.

5. There are wide variations in the population densities of the crown-of-thorns starfish on coral reefs throughout the Indo-Pacific area. It has been impossible to determine whether the high density of *A. planci* in some areas is a unique or a cyclical phenomenon. The evidence is in favour of it being an episodic event, which may have occurred previously.

6. Reef-building is a continuous process of growth and destruction of organisms, including coral and algae. Consolidation of the dead material in the reef mass provides the platform necessary for the continuation of this process. In this context, the feeding by the crown-of-thorns starfish on living coral constitutes, in the long term, a portion of the destruction by natural forces within the reef-building processes.

7. Tourist activity on the Great Barrier Reef has not declined as a result of damage to coral by the starfish.

8. The weights of commercial fish landed in the Cairns–Townsville region have not changed significantly during the past five years.

9. The Committee cannot make any finding on the cause of the population increases in some areas. Of the several theories advanced, two received particular attention. The hypothesis that local collecting of triton shells (*Charonia tritonis*) has reduced predator pressure of the crown-of-thorns starfish has not been substantiated. With regard to the pollution hypothesis, only trace amounts of pesticides and other organic chemicals were found in the tissues of *A. planci* and other organisms collected from infested parts of the Great Barrier Reef. These observation do not support the theory that these substances are the cause of the apparent increases in numbers of the starfish.

10. Any attempt to reduce the population of *A. planci* throughout the whole of the Great Barrier Reef is unwarranted at the present time.

11. Certain reefs, or portions of reefs, have social or commercial importance. In these situations manual destruction of accessible starfish may be feasible.

12. The Committee is of the opinion that knowledge of reef ecology is inadequate to permit a complete assessment of present and future problems concerning the crown-of-thorns starfish and related matters. In view of the unique importance of the Great Barrier Reef there is, therefore, an urgent need for more research. Investigations should include: (i) continued monitor-

ing of *A. planci* populations; (ii) reef ecology, with particular attention to the biology of *A. planci* and corals; (iii) experiments in local control of *A. planci*.

13. It is recommended that a trust fund be established by the Commonwealth and Queensland Governments to provide for research and that an Advisory Committee be appointed to recommend allocations from this fund.

14. It is recommended that a sum of money in the order of $90–$120 thousand be provided in the first year of the Research Trust Fund and that the sum provided be increased progressively by $20,000 in each of the second and third years.

17. James, *Requiem for the Reef,* 67; Robert Raymond, *Starfish Wars. Coral Death and the Crown of Thorns* (Melbourne: MacMillan, 1986), 88.

18. Raymond, *Starfish Wars,* 88.

19. Frank H. Talbot and M. Suzette Talbot, "The Crown-of-thorns Starfish (*Acanthaster*) and the Great Barrier Reef," *Endeavour* 30 (1971):38–42.

20. F. H. Talbot, "Comment," *Search* 2 (1971):192–93.

21. Ibid., 193.

22. P. D. Dwyer, "Crown of Thorns Report," *Search* 2 (1971):361–62, 362.

23. Actually, the committee did rationalize its recommendation for funds for research on political grounds of appeasing concerned citizens: further "population increases" that might occur from time to time would "be viewed with more circumspection and with less public alarm only if there is greater knowledge and understanding of all aspects of the problem." Walsh et al., "Report of the Committee," 24.

24. Dwyer, "Crown of Thorns Report," 362.

25. Ibid., 263.

26. Robert Endean, "Criticism of the Report of the Committee on the Problem of the Crown-of-thorns Starfish (*Acanthaster planci*)," *Operculum* (May–June 1971):36–48.

27. Ibid., 37.

28. Ibid.

29. Ibid., 38.

30. Ibid., 39.

31. Ibid.

32. Ibid., 44.

33. Ibid., 46.

34. Ibid.

35. Ibid., 47.

36. Ibid., 48.

37. James, *Requiem for the Reef,* 19.

38. Ibid., 48.

39. Ibid., 36.

40. Ibid., 41.

41. Ibid.

42. See Rachel Carson, *Silent Spring* (New York: Houghton Mifflin Co., 1962), 83; Charles Elton, *The Ecology of Invasions by Animals and Plants* (London: Methuen and Co., 1957), 15, 131, 148.

43. James reported that by the mid-1970s, the Queensland government had only one or two persons actively working and the expenditure was small, at most around $100,000, *Requiem for the Reef,* 69.

44. Ibid., 1.

Chapter 6

1. J. Y. Cousteau, *The Silent World* (London: Hamish Hamilton, 1953).

2. Rachel Carson, *The Sea Around Us* (New York: Oxford University Press, 1951).

3. Sue Wells, "A Future for Coral Reefs," *New Scientist* 30 (1986):46–50; 47.

4. Ira Rubinoff, Assistant Director (Science), STRI to David Challinor, Acting Assistant Secretary (Science), Smithsonian Institution, March 19, 1971, *Draft: Proposal for Investigations of Acanthaster-Coral Reef interactions in the Eastern Pacific,* David Pawson Papers, Department of Invertebrate Zoology, National Museum of Natural History, Smithsonian Institution, Washington, D.C. On May 20, 1970, David Pawson had prepared the script for a new "*Acanthaster planci* exhibit" at the Museum of Natural History; it included the statement: "The reasons for such a population explosion are unknown. Some biologists believe that if the starfish remains unchecked the very existence of coral reefs in the Pacific and Indian Oceans is threatened."

5. Anthony Cory-Wright to W. Aron, Smithsonian Institution, May 2, 1970. David Pawson Papers.

6. Ibid. See also A. C. Campbell, *Cambridge Red Sea Expedition, 1968 General Report,* February 1969, unpublished, David Pawson Papers; Andrew C. Campbell and Rupert F. G. Ormond, "The Threat of the 'Crown-of-thorns' Starfish (*Acanthaster planci*) to Coral Reefs in the Indo-Pacific Area: Observations on a Normal Population in the Red Sea," *Biological Conservation* 2 (1970): 246–52.

7. Campbell and Ormond, "The Threat of the 'Crown-of-thorns' Starfish (*Acanthaster planci*) to Coral Reefs in the Indo-Pacific Area."

8. William A. Newman, "*Acanthaster:* A Disaster?" *Science* 167 (1970): 1274–75.

9. William Newman, e-mail to the author, August 5, 1996.

10. See, for example, F. P. Shepard, J. R. Curray, W. A. Newman, A. L. Bloom, N. D. Newell, J. I. Tracey and H. H. Veeh, "Holocene Changes in Sea Level: Evidence in Micronesia," *Science* 157 (1967):542–44. A. Ross and W. A. Newman, "A Coral-Eating Barnacle," *Pacific Science* 23 (1969):252–56.

11. William A. Newman, "The Paucity of Intertidal Barnacles in the Tropical Western Pacific," *The Veliger* 2 (1960):89–93; 89.

12. See W. A. Newman, "Darwin and Cirripedology," in J. Truesdale, ed., *The History of Carcinology: Crustacean Issues* 8 (1993):349–34.

13. Ibid.

14. Newman, "*Acanthaster:* A Disaster?," 1274.

15. Ibid.

16. C. H. Edmondson, *Reef and Shore Fauna of Hawaii* (Honolulu: Bernice P. Bishop Museum, Publication No. 22.1946), 295 pp.; 73.

17. Newman, "*Acanthaster:* A Disaster?"

18. Ibid., 1274.

19. Ibid.

20. R. H. Chesher, "*Acanthaster:* A Disaster?" *Science* 167 (1970):1275.

21. Ibid.

22. Ibid.

23. Ibid.

24. Thomas F. Dana, "*Acanthaster:* A Rarity in the Past?" *Science* 169 (1970):894.

25. J. S. Domantay and H. A. Roxas, "The Littoral Asteroidea of Port Galera Bay and Adjacent Waters," *Philippine Journal of Science* 65 (1938):203–38.

26. T. H. Mortensen, "Contributions to the Study of the Development and Larval Forms of Echinoderms I and II," *Mem. de l'Acad. Roy. Sci. et Let., Denmark, Copenhagen, Sec. 9,* 4 (1):1–39.

27. R. Hayashi, "Sea Stars of the Caroline Islands," *Palao Tropical Biological Station Studies* 1 (1938):417–46.

28. Ibid.

29. Masashi Yamaguchi, "*Acanthaster planci* Infestations of Reefs and Coral Assemblages in Japan: A Retrospective Analysis of Control Efforts," *Coral Reefs* 5 (1986):23–30; 23.

30. J. E. Randall, "Chemical Pollution in the Sea and the Crown-of-Thorns Starfish (*Acanthaster planci*)," *Biotropica* 4 (1972):132– 44; 136.

31. Ibid.

32. J. M. Branham, S. A. Reed, Julie H. Bailey, and J. Caperon, "Coral-Eating Sea Stars *Acanthaster planci* in Hawaii," *Science* 172 (1971): 1155–57.

33. Ibid., 1155.

34. Stephen Smith, interview, University of Hawaii, April 19, 1995.

35. Ibid.

36. Peter J. Vine, "Field and Laboratory Observations of the Crown-of-Thorns Starfish, *Acanthaster planci*," *Nature* 228 (1970):341–42; 342.

37. Ibid.

38. Peter J. Vine, "Crown of Thorns (*Acanthaster planci*) Plagues: The Natural Causes Theory," *Atoll Research Bulletin* 166 (1973):1–10; 4.

39. Ibid., 5.

40. John Lucas, "Reproductive and Larval Biology of *Acanthaster planci*

(L.) in Great Barrier Reef Waters," *Micronesica* 9 (1973):197–203.

41. Vine, "Crown of Thorns (*Acanthaster planci*) Plagues," 6.

42. Ibid., 7.

43. James W. Porter, "Predation by *Acanthaster* and its Effect on Coral Species Diversity," *The American Naturalist* 106 (1972):487–92; 487. Porter cited several papers in support: D. J. Barnes, R. W. Brauer, and M. R. Jordan, "Locomotory Responses of *Acanthaster planci* to Various Species of Coral," *Nature* 228 (1970):342–44; J. M. Branham, S. A. Reed, J. H. Bailey, and J. Caperon, "Coral Eating Sea Stars *Acanthaster planci* in Hawaii, " *Science* 172 (1971):115–17; T. F. Goreau, J. C. Lang, E. A. Graham, and P. D. Goreau, "Structure and Ecology of Saipan Reefs in Relation to Predation by *Acanthaster planci* (Linnaeus)," *Bulletin of Marine Science* 22 (1972):13–152.

44. R. T. Paine, "Food Web Complexity and Species Diversity," *American Naturalist* 100 (1966):65–75.

45. Paine suggested that the giant triton might play the same role as *Pisaster*, if it actually kept *Acanthaster* in check. If this proved not to be the case, he suggested that *Acanthaster* itself might be comparable to *Pisaster*. *Acanthaster* would control coral, the prime monopolist of space on reefs, in the same way that *Pisaster* was capable of controlling mussels. See Robert Paine, "A Note on Trophic Complexity and Community Stability," *American Naturalist* 103 (1969):91–93.

46. Porter, "Predation by *Acanthaster*," 490.

47. M. D. Bellamy, "Crown of Thorns, a Jekyll-Hyde Starfish," *Marine Aquarist* 6 (1975):49–58; 49.

48. J. M. Branham, "The Crown of Thorns on Coral Reefs," *BioScience* 23 (1973):219–26; 219.

49. Ibid., 221.

50. Ibid., 225.

51. Ibid., 223.

52. Ibid., 224.

53. Ibid.

54. Ibid.

55. Charles Darwin, *1860. The Voyage of the Beagle*. Natural History Library Edition. (New York: Anchor Books, Double Day, 1962), 459.

56. Ibid., 476.

57. Charles Darwin, *1842. On the Structure and Distribution of Coral Reefs* (Felling-on-Tyne: Walter Scott, Publishing, 1905), 30–31.

58. Branham, "The Crown of Thorns on Coral Reefs," 225.

59. Ibid.

60. Thomas F. Dana, William A. Newman, and Edward W. Fager, "*Acanthaster* Aggregations: Interpreted as Primarily Responses to Natural Phenomena," *Pacific Science* 26 (1972):355–72; 362.

61. Ibid., 362.

62. Ibid., 370.

63. Ibid., 366.

64. Ibid.

65. Ibid., 369.

66. Ibid.

67. Ibid., 368.

68. Ibid., 370.

69. Ibid.

70. Ibid., 367.

71. Ibid., 370.

72. Ibid., 385.

73. Branham, "The Crown of Thorns on Coral Reefs," 222.

74. Anon., "Threat of Starfish to Coral Ends," *New York Times*, January 4, 1972, p. 27, C-1.

75. Nierenberg's speech was reported in *Sports Fishing Institute Bulletin* No. 245, June 1973. See also anon., "The Non-problem," *Sea Winds* 17 No. 2 (1973):6. It was also reported by Robert Raymond, *Starfish Wars. Coral Death and the Crown-of-thorns* (Melbourne: MacMillan, 1986), 91–92.

Chapter 7

1. R. E. Johannes, "Pollution and Degradation of Coral Reef Communities," in E. J. Ferguson Wood and R. E. Johannes, eds., *Tropical Marine Pollution* (Amsterdam: Elsevier Scientific Publishing, 1975), 13–51; 51.

2. Stephen Smith, Interview, Department of Oceanography, University of Hawaii, Honolulu, April 19, 1995.

3. Ibid.

4. James D. Watson, *The Double Helix* (New York: Antheneum, 1968).

5. Charles Darwin (1860), *The Voyage of the Beagle* (London: Dent, 1959).

6. Paul R. Ehrlich, *The Population Bomb* (Rivercity, Mass: Rivercity Press, 1975).

7. Thomas Malthus, 1798. *An Essay on the Principle of Population, As It Affects the Future Improvement of Society, with Remarks on the Speculations of Mr. Godwin, M. Condorcet, and Other Writers*. J. Johnson. Reprint. (Ann Arbor: University of Michigan Press, 1959).

8. See, for example, Robert C. Paehlke, *Environmentalism and the Future of Progressive Politics* (New Haven: Yale University Press, 1989), 41–75. Garrett Hardin, *Living Within Limits: Ecology, Economics and Population Taboos* (New York: Oxford University Press, 1993).

9. John Ogden, interview, Florida Institute for Marine Science, St. Petersburg, March 16, 1994.

10. Ibid.

11. John Ogden, Richard Brown and Norman Salesky, "Grazing by the Echinoid *Diadema antillarum Phillippi*: Formation of Halos around the West Indian Patch Reefs," *Science* 182 (1973):715–17.

12. See, for example, Leslie Roberts, "Zebra Mussel Invasion Threatens U.S. Water," *Science* 249 (1990):1370–72; Brian Banks, "Alien Onslaught," *Equinox* No. 53 (1990):69–75; Tim Walker, "Dreissena Disaster," *Science News* 139(1991): 282–84; Michael Ludyanskiy et al., "Impact of the Zebra Mussel, a Bivalve Invader," *BioScience* 43 (1993):533–41.

13. H. A. Lessios, "Mass Mortality of *Diadema antillarum* in the Caribbean: What Have We Learned?" *Annual Review of Ecological Systems* 19 (1988): 371–93. Monitoring its populations over the subsequent ten years, biologists showed that its recovery was very slow, and how its near extinction had both immediate and long-term effects on the distribution and abundance of other species members of the coral-reef communities. See H. A. Lessios, "*Diadema antillarum* 10 Years after Mass Mortality: Still Rare, Despite Help from a Competitor," *Proceedings of the Royal Society of London* B 259 (1995):331–37.

14. Ogden, interview.

15. Robert Johannes, interview, Panamá, June 26, 1996.

16. Ibid.

17. Frank Benjamin Golley, *A History of the Ecosystem Concept in Ecology* (New Haven: Yale University Press, 1993); Donald Worster, *Nature's Economy: A History of Ecological Ideas* (Cambridge: Cambridge University Press, 1977).

18. A. G. Tansley, "The Use and Abuse of Vegetational Concepts and Terms," *Ecology* 16 (1935):284–307. The arrival of the new ecology was marked in 1942 with the appearance of Raymond Lindeman's paradigmatic paper on "The Trophic-Dynamic Aspect of Ecology." He studied Cedar Bog Lake with the aim of quantifying the productivity and efficiency of the food chain: how energy is passed on from one trophic level (feeding level) to the next. One of the most salient facts was that energy in use at one trophic level could never be fully passed on to the next higher level up the food chain. Lindeman aimed to calculate the productivity of each level in the food chain and the efficiency of energy transfers. R. L. Lindeman, "The Tropic-Dynamic Aspect of Ecology," *Ecology* 23 (1942):399–408.

19. For general historical overviews of the revolutionary ideas and approaches of this period, see Worster, *Nature's Economy*; Peter J. Bowler, *The Environmental Sciences* (New York: Norton, 1992).

20. Eugene Odum, *Fundamentals of Ecology* (Philadelphia: W. B. Saunders, 1959).

21. At that time, it was generally believed that bacteria were the most important nutrient regenerators in the oceans because they are in the soil and on land. He and his colleague Lawrence Pomeroy took an alternative perspective, generally accepted today, that bacteria were actually competitors for dissolved nutrients such as nitrogen and phosphorus, and that it was "small animals" that were doing most of the nutrient regeneration. They worked on protozoa and small zooplankton and measured excretion rates both in salt

marshes and out in boats studying zooplankton metabolism.

22. Howard Odum and Eugene Odum, "Trophic Structure and Productivity of a Windward Coral Reef Community on Eniwetok Atoll," *Ecological Monographs* 25 (1955):291–320.

23. For a general discussion of symbiosis of coral and algae, see Angela E. Douglas, *Symbiotic Interactions* (Oxford: Oxford University Press, 1994). See also L. Muscatine, "The Role of Symbiotic Algae in Carbon and Energy Flux in Reef Corals," *Coral Reefs* 10 (1990):75–87.

24. Smith, interview.

25. Promoted, during the 1960s and early 1970s, by such organizations as the International Biology Programme, ecosystem ecologists asked questions such as what is the natural energy cost of agriculture and how efficient is it, or how can one use the geographical variability and the biochemical activity of plants and animals to contribute to economic developments of some countries? The IBP placed great importance on the discovery of plants, wild or cultivated, capable of maximum photosynthesis per unit area of foliage, or ground surface, or capable of maximum production of those proteins or amino acids essential to humans. Special emphasis was on those processes having a key position in ecosystem productivity: the use of solar energy by plants (photosynthesis) and the fixation of molecular nitrogen by organisms (chemosynthesis). See E. B. Worthington, *The Evolution of the IBP*, 26, 261.

26. R. E. Johannes, "Introduction," *Energy and Nutrient Flux in D. R. Stoddart and R. E. Johannes, Coral Reefs: Research Methods* (Paris: UNESCO, 1978), 349–51.

27. Ibid., 349, 350.

28. "At the outset of Project Symbios," Johannes recalled, "some of us were also optimistic that a better understanding of energy and nutrient flux in reef communities would provide us with some insight into the problem of harvesting edible reef organisms efficiently." Ibid., 350.

29. R. E. Johannes, *Words of the Lagoon: Fishing and Marine Lore in the Palau District of Micronesia* (Berkeley: Univ. of California Press, 1981), x.

30. Ibid.

31. R. E. Johannes, "Pacific Island Peoples' Science and Marine Resource Management," in John Morrison, Paul Geraghty, Linda Crowl, eds., *Science of Pacific Island Peoples* (Suva, Fiji: Institute of Pacific Studies, The University of the South Pacific, 1994), 81–89. Johannes, *Words of the Lagoon*.

32. R. E. Johannes, "Traditional Marine Conservation Methods in Oceania and their Demise," *Annual Review of Ecology and Systematics* 9 (1978):349–64.

33. Garrett Hardin, "The Tragedy of the Commons," *Science* 16 (1968): 1243–48.

34. Y. I. Sorokin, *Coral Reef Ecology: Ecological Studies* 102 (Berlin: Springer-Verlag, 1993), 421.

35. See R. E. Johannes and W. MacFarlane, *Traditional Fishing in the*

Torres Strait Islands (Hobart: CSIRO, 1991).

36. Johannes, interview, 1996.

37. R. E. Johannes, "How to Kill a Coral Reef," *Marine Pollution Bulletin* 2 (1971):9–10.

38. It is accepted today that zooplankton are important for providing nitrogen and phosphorous to the corals, but in terms of energy, the bulk of it in reef-building corals comes from their symbiotic algae.

39. R. E. Johannes, "Pollution and Degradation of Coral Reef Communities," 51.

40. Ibid.

41. Ibid.

42. Ibid., 42.

43. Johannes, "How to Kill a Coral Reef," 10.

44. Richard Chesher, to the author, February 3, 1998.

45. Ibid.

46. Robert Endean and Richard H. Chesher, "Temporal and Spatial Distribution of *Acanthaster planci* Population Explosions in the Indo-West Pacific Region," *Biological Conservation* 5 (1973):87–95.

47. T. P. Bligh and N. Bligh, "*Acanthaster* in the Indian Ocean" *Nature* 229 (1971):281.

48. D. Owens, "*Acanthaster planci* starfish in Fiji: Survey of Incidence and Biological Studies," *Fiji Agricultural Journal* 33 (1971):15–23.

49. M. Nishihira and K. Yamazato, "Brief Survey of *Acanthaster planci* in Sesoko Island and its Vicinity, Okinawa," Technical Report, Sesoko Marine Science Laboratory, University of the Ryukyus 1 (1972): 1–20. Cited in J. Randall, "Chemical Pollution in the Sea and the Crown-of-Thorns Starfish, *Acanthaster planci*," *Biotropica* 4 (1972);132–44.

50. Ibid., 136.

51. R. T. Tsuda, "Status of *Acanthaster planci* and Coral Reefs in the Mariana and Caroline islands, June 1970 to May 1971." Marine Laboratory. University of Guam, Technical Report 2, 1971. See also, D. P. Cheney, "Guam and the Crown-of-thorns Starfish," *Guam Rec.* 1 2/3 (1972):74–80.

52. Randall, "Chemical Pollution in the Sea," 134. They visited Tahiti, Moorea, Bora Bora, and Huahine (Society Islands); Anuanuraro, Mangereva, Temoe, Rangiroa, Manihiki, and Takakaroa (Tuamotu Archipelago); Pitcairn, Oeno, Henderson, and Ducie (Pitcairn Group); Rapa; Raivavae, Tubuai, and Rurutu (Austral islands); Rarotonga (Cook Islands).

53. Ibid., 138.

54. Chesher and Endean, "Temporal and Spatial Distribution of *Acanthaster planci*," 93.

55. Ibid., 87.

56. See Randall, "Chemical Pollution in the Sea," 137.

57. J. L. Fischer, "Starfish Infestation: Hypothesis," *Science* 165 (1969):645.

58. Randall, "Chemical Pollution in the Sea," 134.

59. Ibid., 138.

60. Ibid., 139.

61. Randall, cited in P. A. Butler "Commercial Fisheries Investigations," in *Pesticide-wildlife Studies*, U.S. Department of the Interior. Circulation Fish Wildlife Service 167 (1963):11–25.

62. Randall, 139. It had only recently become apparent that the compounds known as polychlorinated hydrocarbons (PCBs), were almost as widespread as DDT in the global environment. PCBs had been used in the manufacture of such products as plastics, paints, and hydraulic fluids long before DDT was discovered. Swedish scientists found these compounds along with DDT in mussels, various fishes, seals, and sea birds in Swedish waters. S. Jensen, A. G. Johnels, M. Olsson, and G. Otterlind, "DDT and PCB in Marine Animals from Swedish Waters," *Nature* 224 (1969):247–50. But Randall suspected that they were an unlikely suspect in the *Acanthaster* population explosions. It seemed that they would require a greater concentration to be toxic than chlorinated hydrocarbon pesticides. Moreover, many of the areas where the starfish was very abundant had little or no industrial pollution.

63. W. J. L. Sladen, C. M. Manzie, and W. L. Reichel, "DDT Residues in Adelie Penguins and a Crabeater Seal from Antartica: Ecological Implications," *Nature* 210 (1966):670–73.

64. Randall, "Chemical Pollution in the Sea," 135.

65. C. F. Wurster "DDT Reduces Photosynthesis by Marine Phytoplankton," *Science* 159 (1968):1474–75.

66. Randall, "Chemical Pollution in the Sea," 140.

67. Ibid., 142.

68. Peter Glynn, interview with Rosentiel School of Marine and Atmospheric Science, University of Miami, March 14, 1994.

69. Peter Glynn, interview, Rosentiel School of Marine and Atmospheric *Science*, March 12, 1997.

70. Ibid.

71. Peter Glynn, "*Acanthaster*: Effect on Coral Reef Growth in Panama," *Science* 180 (1973):504–06.

72. Peter Glynn, "The Impact of *Acanthaster* on Corals and Coral Reefs in the Eastern Pacific," *Environmental Conservation* 1 (1974):295–304; 295.

73. W. A. Newman and T. F. Dana, "*Acanthaster*: Test of the Time Course of Coral Destruction, " *Science* 183 (1974):103.

74. Glynn, "The Impact of *Acanthaster* on Corals," 300–301.

75. Ibid., 301.

76. Ibid.

77. Ibid.

78. W. Wickler, "Biology of *Hymenocera* picta Dana," *Micronesica* 9 (1973): 225–30.

79. R. F. G. Ormond and A. C. Campbell, "Formation and Breakdown of

Acanthaster planci Aggregations in the Red Sea," *Proceedings of the Second International Coral Reef Symposium* 1 (1974):595–619.

80. Peter Glynn, "Some Physical and Biological Determinants of Coral Community Structure in the Eastern Pacific," *Ecological Monographs* 46 (1976):431–56. There is a minor drama related to the publication of this refutation of James Porter's hypothesis regarding *acanthaster's* prey preference for *Pocillopora* and its effect on coral diversity. James Porter was editor of *Ecological Monographs* when Glynn's paper, highly critical of his hypothesis, was received. Porter explains: "I can assure you that no paper ever submitted is perfect, and an experienced editor has considerable power over the fate of a submitted manuscript. By choosing reviewers who are known to be hypercritical, or simply by footdragging during the inherently lengthy manuscript review process, I was in a position either to significantly delay its publication or to steer the paper into a rejection. Instead, I did what I was supposed to do and made sure it was published promptly. I have never made much of 'doing the right thing.' On the other hand, we hear so much in science about the malfeasance, competition, and self-promotion, that a relevant counter example in the context of the *Acanthaster* controversy might be of interest to your readers." James Porter, letter to the author, January 21, 1997.

81. Peter Glynn, "Interaction Between *Acanthaster* and *Hymenocera* in the Field and the Laboratory," *Proceedings of the Third International Coral Reef Symposium* 1977, 209–15.

82. Ibid., 212.

83. Ibid., 213–14.

84. Ibid., 214–15. Glynn also pointed out that Richard Randall provided evidence of low mortality of such guarded corals following *Acanthaster* outbreaks at Guam. R. H. Randall, "Distribution of Corals after *Acanthaster planci* (L.) Infestations at Tanguisson Point, Guam," *Micronesica* 9 (1973):213–22.

85. Peter Glynn, "Some Ecological Consequences of Coral-Crustacean Guard Mutualisms in the Indian and Pacific Oceans," *Symbiosis* 4 (1987):301–24. Richard Randall also provided evidence of low mortality of such guarded corals following Acanthaster outbreaks at Guam. R. H. Randall, "Distribution of Corals After *Acanthaster planci* (L.) Infestations at Tanguisson Point, Guam," 213–22.

Chapter 8

1. See Marshall Logan, *The Story of the Panama Canal* (Philadelphia: J. C. Winston, 1913). David G. McCullough, *The Path Between the Seas: The Creation of the Panama Canal, 1870–1914* (New York: Simon and Schuster, 1977); Walter LaFeber, *The Panama Canal: The Crisis in Historical Perspective* (New York: Oxford University Press, 1989).

2. See David Challinor, "Background for a New, Sea-Level, Panama Canal," in M. L. Jones, ed., *The Panama Biota: Some Observations Prior to A*

Sea-Level Canal, Bulletin of the Biological Society of Washington, no. 2, 1972, 7–12.

3. Ibid., 10.

4. Marti Mueller, "New Canal: What about Bioenvironmental Research?" *Science* 163 (1969): 165–67.

5. Col. Harold G. Stacy, "Medical Support of Studies for a Sea Level Canal," *Military Medicine* 134 (1969):1355–62.

6. Ibid.

7. Challinor, "Background for a New, Sea-Level, Panama Canal," 10.

8. Ibid.

9. Ibid., 8.

10. See LaMont C. Cole, "Can the World be Saved?" *BioScience* 18(1968): 679–84; 682. See also William E. Martin, "Bioenvironmental Studies of the Radiological-Safety Feasibility of Nuclear Excavation," *Bioscience* 19 (1969): 135–37; James R. Vogt, "Radionuclide Production for the Nuclear Excavation of an Isthmian Canal," *BioScience* 19 (1969):138–39; R. L. Charnell, T. M. Zorich, and D. E. Holly, "Hydrologic Redistribution of Radionuclides around a Nuclear-Excavated Sea-Level Canal," *BioScience* 19 (1969): 799–803; Col. Ernest Graves, Jr., "Nuclear Excavation of an Isthmian, Sea-Level Canal," *SAE Transactions* 75 (1965):93–108. Lt. Col. Bernard C. Hughes, "Nuclear Excavation Design of a Transmisthmian Sea-Level Canal," *Nuclear Applications and Technology* 7 (1969):305–27.

11. William A. Newman, "The National Academy of Science Committee on the Ecology of the Interoceanic Canal," in M. L. Jones, ed., *The Panamic Biota: Some Observations Prior to a Sea-level Canal, Bulletin of the Biological Society of Washington* No. 2 (1972):247–59; 249.

12. Robert W. Topp, "Interoceanic Sea-Level Canal: Effects on the Fish Faunas," *Science* 163 (1969):1324–27.

13. Challinor, "Background for a New, Sea-Level, Panama Canal," 11.

14. Ernst Mayr, interview, Cambridge, Mass., November 21, 1996.

15. Wolfgang Saxon, "Martin H. Moynihan, 68, an Authority on Animal Behavior," *New York Times*, Obituaries, Sunday December 15, 1996, 67.

16. Ibid.

17. Ira Rubinoff, "Central American Sea-Level Canal: Possible Biological Effects," *Science* 161 (1968):857–61; 860. See also Christopher Weathersbee, "Linking the Oceans," *Science News* 94 (1968):578–81.

18. Rubinoff, "Central American Sea-Level Canal," 861.

19. John C. Briggs, "The Sea-Level Panama Canal: Potential Biological Catastrophe," *BioScience* 19 (1969):44–47; 46.

20. John C. Briggs, "Panama's Sea-Level Canal," *Science* 162 (1968):511-13.

21. Gairdener B. Moment, "A Disaster?" *BioScience* 19 (1969):497.

22. Marti Mueller, "New Canal: What about Bioenvironmental Research?," *Science* 163 (1969): 165–67; 167.

23. John P. Sheffey, "When Caribbean and Pacific Waters Mix," *Science* 162 (1968):1329.

24. See also John P. Sheffey, "Unnecessary Alarm," *BioScience* 19 (1969):300–01.

25. See for example, John C. Briggs "Briggs' Reply," *BioScience* 19 (1969): 301.

26. Mueller, "New Canal: What about Bioenvironmental Research?" 167.

27. Joshua Lederberg, "Sea-Level Canal Points Up Need for Environmental Data," *Washington Post*, February 1, 1969.

28. Ogden, Interview, Florida Institute for Marine Science, St. Petersburg, March 16, 1995.

29. Glynn, "The Impact of *Acanthaster* on Corals," 298.

30. R. E. Johannes, "How to Kill a Coral Reef," *Marine Pollution Bulletin* 2 (1971):9–10.

31. Mayr, interview.

32. Newman, "The National Academy of Science Committee," 249.

33. Ibid., 357.

34. Ibid. One can add to this the zebra mussel outbreaks in the Great Lakes and other inland waters that began in the mid-1980s, believed to have been transferred from the Caspian Sea in the ballast water of a transatlantic freighter. See, for example, Leslie Roberts, "Zebra Mussel Invasion Threatens U.S. Water," *Science* 249 (1990):1370–72; Brian Banks, "Alien Onslaught," *Equinox* No. 53 (1990):69–75; Tim Walker, "Dreissena Disaster," *Science News* 139(1991):282–84; Michael Ludyanskiy et al., "Impact of the Zebra Mussel, a Bivalve Invader," *BioScience* 43 (1993):533–41.

35. Newman, "The National Academy of Science Committee," 257.

36. Ibid., 251.

37. Ibid., 252.

38. Ibid., 256.

39. Ibid.

40. Ibid.

41. See I. Rubinoff and C. Kropach, "Differential Reactions of Atlantic and Pacific Predators to Sea Snakes," *Nature* 228 (1970):1288–90.

42. William Newman, e-mail to the author, October 18, 1996.

43. Ibid.

44. William Newman, e-mail to the author, August 5, 1996.

45. W. A. Newman and T. F. Dana, "*Acanthaster*: Test of the Time Course of Coral Destruction," *Science* 183 (1974):103.

46. Philip M. Boffey, "Sea-Level Canal: How the Academy's Voice was Muted," *Science* 171 (1971):355–58; 355.

47. Challinor, "Background for a New, Sea-Level, Panama Canal," 9.

48. Boffey, "Sea-Level Canal," 355.

49. A limited number of copies of the report and supporting annexes, entitled, "Interoceanic Canal Studies 1970," were made available without

charge from the Office of the Deputy Undersecretary of the Army, Washington, D.C. 20310.

50. Boffey, "Sea-Level Canal," 355.

51. Ibid.

52. Ibid., 357.

53. Ibid.

54. Ibid.

55. Ibid., 358.

56. Challinor, "Background for a New, Sea-Level, Panama Canal," 9.

57. Ogden, interview.

Chapter 9

1. R. J. Walsh, J. M. Harvey, W. G. H. Maxwell, and J. M. Thomson, "Report on Research Sponsored by the Advisory Committee on Research into the Crown of Thorns Starfish," Australian Government Publishing Service, 15 October 1975, 3.

2. Applications were examined by external assessors and the Advisory Committee made final recommendations to the Australian Minister for Science, who, in turn, submitted them to the Queensland Minister for Primary Industries. After both ministers approved them, a press announcement was released. Ibid., 6.

3. Ibid.

4. Ibid., 3.

5. John Lucas, e-mail to the author, December 12, 1996.

6. Ibid.

7. J. S. Lucas, "Reproductive and Larval Biology of *Acanthaster planci* (L.) in Great Barrier Reef Waters," *Micronesica* 9 (1973):197–203; Idem, "Environmental Influences on the Early Development of *Acanthaster planci* (L.)," in *Crown-of-Thorns Starfish Seminar Proceedings, Brisbane, 6 September 1974* (Canberra: Australian Government Publishing Service, 1975), 103–21.

8. Walsh et al., "Report on Research Sponsored by the Advisory Committee," 14. R. G. Pearson, "Coral Reefs, Unpredictable Climatic Factors and *Acanthaster,*" in *Crown-of-Thorns Starfish Seminar Proceedings, Brisbane, 6 September 1974* (Canberra: Australian Government Publishing Service, 1975), 131–34.

9. Walsh et al., "Report on Research Sponsored by the Advisory Committee," 17.

10. Ibid., 15.

11. Ibid., 16.

12. E. Frankel, "*Acanthaster* in the past: Evidence from the Great Barrier Reef," *Crown-of-Thorns Starfish Seminar Proceedings, Brisbane, 6 September 1974* (Canberra: Australian Government Publishing Service, Australia, 1975), 159–66; Idem, "Previous *Acanthaster* Aggregations in the Great Barrier Reef," *Proceedings of the 3rd International Coral Reef Symposium*

(1977):201–08.

13. Walsh et al., "Report on Research Sponsored by the Advisory Committee," 19.

14. Ibid., 25.

15. Ibid., xii.

16. Ibid., 30.

17. R. Endean and W. Stablum, "A Study of Some Aspects of the Crown-of-Thorns Starfish (*Acanthaster planci*) Infestations of Reefs of Australia's Great Barrier Reef," *Atoll Research Bulletin*, No. 167, November 23 (1973):1–72, 1.

18. Walsh et al., "Report on Research Sponsored by the Advisory Committee," 19.

19. Ibid., 7.

20. Ibid., 10.

21. Richard Kenchington, "The Crown-of-thorns Crisis in Australia: A Retrospective Analysis," *Environmental Conservation* 5 (1978): 11–20; 17.

22. Richard Kenchington, e-mail to the author, February 8, 1997.

23. R. A. Kenchington, "*Acanthaster planci* on the Great Barrier Reef: Detailed Surveys of Four Transects Between 19 and 20 S," *Biological Conservation* 9 (1976):165–79.

24. Walsh et al., "Report on Research Sponsored by the Advisory Committee," 10.

25. Ibid., 27.

26. Ibid., 26.

27. Ibid.

28. Ibid.

29. Charles Birkeland and John S. Lucas, *Acanthaster planci: Major Management Problem of Coral Reefs* (Boca Raton: CRC Press, 1990), 10.

30. Richard Kenchington, "*Acanthaster Planci* and Management of the Great Barrier Reef," *Bulletin of Marine Science* 41 (1987):522–60.

31. Richard Kenchington, "The Crown-of-thorns Crisis in Australia, 11–20.

32. Ibid., 19.

33. Ibid., 18.

34. Ibid., 19.

35. Ibid., 18.

36. Ibid.

37. Ibid.

38. Ibid.

39. Ibid., 13–14. Lu Eldredge recalls that as chairman of the Biology Department at the University of Guam, he received a phone call one day from Vancouver: "Some talk show wanted to interview me about the *Acanthaster* thing. And they said, "Is the island going to topple over? Is there going to be a problem with the island? Are they going to come across the pacific and start

attacking the west coast?" Government officials were no better; Eldredge rec-ollects: "There is a legend, a myth on Guam itself that Guam is built on three pillars under water and in the legislature one of our Senators was con-cerned that *Acanthaster* was going to eat through the pillars and the island would topple over." And this was a serious statement said in public on the floor of the legislature." Lu Eldredge, interview, Honolulu, April 18, 1995.

40. Kenchington, "The Crown-of-thorns Crisis in Australia," 15.

41. Jon N. Weber, "Disaster at Green Island—Other Pacific Islands May Share Its Fate," *Earth and Mineral Sciences* 38 (1969):37–41; 37.

42. Bernard Dixon, "Doomsday for Coral?" *New Scientist* 44 (1969):226–27; 226.

43. Kenchington, "The Crown-of-thorns Crisis in Australia," 11.

44. Ibid., 16.

45. Ibid.

46. See also R. A. Kenchington and B. Morton, *Two Surveys of the Crown-of-thorns Starfish over a Section of the Great Barrier Reef* (Canberra: Australian Government Publishing Service, 1976).

47. Kenchington, "The Crown-of-thorns Crisis in Australia," 17.

48. Ibid., 17. Kenchington noted that the number of marine biologists working in Australia rose from 73 to 153 between 1963 and 1970.

49. Ibid., 19.

50. See, for example, R. H. Bradbury, "Lessons of the Crown-of-thorns Research," *Search* 7 (1976):461–62.

51. Kenchington, "The Crown-of-thorns Crisis in Australia," 17.

52. Ibid., 17–19.

53. R. G. Pearson, "Recolonization by Hermatypic Corals of Reefs Dam-aged by *Acanthaster*," *Proceedings of the Second International Coral Reef Sympo-sium* (1974):207–16.

54. Kenchington, "The Crown-of-thorns Crisis in Australia," 17.

55. Ibid., 18.

56. David Alexander, "Guarding a Watery Paradise," *International Wildlife* 18, No. 3, (1988):4–10; 6.

57. Graeme Kelleher and Wendy Craik, "Maintaining Biodiversity in a Managed Land and Seascape: The Great Barrier Reef, Australia," in Robert Szaro and David Johnston, eds., *Biodiversity in Managed Landscapes* (New York: Oxford University Press, 1996), 492–505.

58. Ibid. See also Yuri I. Sorokin, *Coral Reef Ecology: Ecological Studies*, vol. 102 (Berlin: Springer-Verlag, 1993).

59. David Alexander, "Guarding a Watery Paradise."

60. Ibid., 8–9.

61. Sorokin, *Coral Reef Ecology*, 421.

62. Kenchington, "The Crown-of-thorns Crisis in Australia," 18.

63. R. A. Kenchington and R. Pearson, "Crown of Thorns Starfish on the Great Barrier Reef: A Situation Report," *Proceedings of the Fourth Interna-*

tional Coral Reef Symposium, Manila, 1981, Vol. 2: 597–600; 599.

64. Robert Raymond, "A Conspiracy of Silence over the Destruction of the Reef," *The Bulletin*, December 18, 1984, 68–77; 68.

65. "Report of Crown of Thorns Starfish Advisory Committee 1980," (April 1984). S. 1. Obtained from Charles Birkeland, University of Guam.

66. Ibid., 2.

67. "Report on Experiments to Assess Eradication Techniques for Use Against *Acanthaster planci*," (April–December 1981) S.4. Obtained from Charles Birkeland.

68. Kenchington and Pearson, "Crown of Thorns Starfish on the Great Barrier Reef: A Situation Report," 599.

69. "Oral History of Great Barrier Reef, Draft Report," by B. J. Dalton, History Department, James Cook University, North Queensland, 1–28. Obtained from Charles Birkeland.

70. Ibid., 3.

71. Ibid., 23.

72. Ibid., 3–4.

73. Ibid., 26.

74. R. Endean, "*Acanthaster planci* Infestation of Reefs of the Great Barrier Reef," *Proceedings Third International Coral Reef Symposium*, May 1977, 185–92.

Chapter 10

1. Frank N. Egerton, "Changing Concepts of the Balance of Nature," *The Quarterly Review of Biology* 48 (1973):322–50; 322.

2. Joseph H. Connell and Wayne P. Sousa, "On the Evidence Needed to Judge Ecological Stability or Persistence," *The American Naturalist* 121 (1983): 789–824; 808.

3. See Egerton, "Changing Concepts of the Balance of Nature."

4. During the late nineteenth century struggle and competition were elaborated equally in the "natural" and human social realms. The use of natural law as the basis for a given view of society became commonplace in social, political, and economic theory. See, for example, John Greene, "Biology and Social Theory in the Nineteenth Century," in Marshall Clagett, ed., *Critical Problems in the History of Science* (Madison: University of Wisconsin Press, 1962), 416–46; Richard Hofstadter, *Social Darwinism in American Thought*, rev. ed. (Boston: Beacon Press, 1955); Greta Jones, *Social Darwinism in English Thought* (London: Harvester, 1980); Robert Young, *Darwin's Metaphor: Nature's Place in Victorian Culture* (Cambridge: Cambridge University Press, 1985); "Darwinism Is Social," in David Kohn, ed., *The Darwinian Heritage* (Princeton: Princeton University Press, 1985), 609–38; Jim Moore, "Socializing Darwinism: Historiography and the Fortunes of a Phrase," in Les Levidow, ed., *Science as Politics* (London: Free Association Books, 1986), 38–75.

5. Camille Limoges, "Milne-Edwards, Darwin, Durkheim and Division of Labor: A Case Study in the Reciprocal Conceptual Exchanges Between the Social and Natural Sciences," in I. B. Cohen, ed., *The Relations Between the Natural Sciences and the Social Sciences* (The Netherlands: Kluwer Academic Publishers, 1994), 317–43.

6. Charles Darwin, *On the Origin of Species by Means of Natural Selection* (London: John Murray, 1859), reprinted with an introduction by Ernst Mayr (Cambridge: Harvard University Press, 1966), 115.

7. Herbert Spencer, *The Principles of Biology*, vol. 2, revised and enlarged (New York: D. Appelton and Co., 1899), 396–408.

8. Ibid., 396.

9. Ibid., 405.

10. Ibid., 408.

11. See Jan Sapp, *Evolution by Association: A History of Symbiosis* (New York: Oxford University Press, 1994).

12. See, for example, Donald Worster, *Nature's Economy: A History of Ecological Ideas* (Cambridge: Cambridge University Press, 1977).

13. Frederick Clements, *Plant Succession: An Analysis of the Development of Vegetation* (Washington: Carnegie Institute of Washington, 1916), 3.

14. Charles E. Elton, *The Ecology of Invasions by Animals and Plants* (London: Methuen and Co., 1957), 145–53.

15. Ibid., 145.

16. Ibid., 147.

17. Ibid., 148.

18. Ibid., 148–49.

19. Ibid., 149.

20. R. H. MacArthur, "Fluctuation of Animal Populations and a Measure of Community Stability," *Ecology* 36 (1955):533–36; 535.

21. G. E. Hutchinson, "Homage to Santa Rosalia or Why are There So Many Kinds of Animals?" *The American Naturalist* 93 (1959):145–59; 156.

22. R. Endean, "*Acanthaster planci* Infestation of Reefs of the Great Barrier Reef," *Proceedings Third International Coral Reef Symposium* 1 (1977): 185–92; 188.

23. Ibid., 189.

24. H. L. Sanders, "Marine Benthic Diversity: A Comparative Study," *American Naturalist* 102 (1968):243–82; J. F. Grassle, "Variety in Coral Reef Communities," in *Biology and Geology of Coral Reefs, Volume II, Biology 1*, O. A. Jones and R. Endean, eds. (New York: Academic Press), 247–70.

25. R. T. Paine, "A Note on Trophic Complexity and Community Stability," *American Naturalist* 103 (1969):91–93; 91.

26. See Nelson G. Hairston, Frederick E. Smith, and Lawrence B. Slobodkin, "Community Structure, Population Control, and Competition," *The American Naturalist* 94 (1960):421–24; William Murdoch, "'Commu-

nity Structure, Population Control and Competition'—A Critique," *American Naturalist* 100 (1966):219–26; Paul Ehrlich and L. C. Birch, "The 'Balance of Nature' and 'Population Control,'" *The American Naturalist* 101 (1967):97–107; L. B. Slobodkin, F. E. Smith and N. G. Hairston, "Regulation in Terrestrial Ecosystems, and the Implied Balance of Nature," *American Naturalist* 101 (1967):109–24.

27. Sanders, "Marine Benthic Diversity: A Comparative Study," 243–82.

28. Stuart L. Pimm, *The Balance of Nature? Ecological Issues in the Conservation of Species and Communities* (Chicago: University of Chicago Press, 1991). See also R. C. Lewontin, "The Meaning of Stability," in G. M. Woodwell and H. H. Smith, eds., *Diversity and Stability in Ecological Systems. Brookhaven Symposium in Biology 22* (Upton, N.Y.: Brookhaven National Library, 1969), 13–24. The relation between complexity and stability was frequently discussed at the First International Congress of Ecology in The Hague, The Netherlands, September 8–14, 1974. See *Proceedings of the First International Congress of Ecology*, Centre for Agricultural Publishing and Documentation, Wageningen, The Netherlands, 1974.

29. R. M. May, "Stability in Multi-species Community Models," *Mathematical Biosciences* 12 (1971):59–79; "Will a Large Complex System be Stable?" *Nature* 238 (1972) 413–14.

30. Robert M. May, *Stability and Complexity in Model Ecosystems* (Princeton: Princeton University Press, 1973); revised 2nd edition 1974. May elaborated on how nonlinearities could produce cyclical oscillations, on the role played by time delays, and the importance of understanding the strength of the interactions between species. One issue was certain for him: species integration "is a very non-linear affair, and complex communities contain much more 'information' than can be estimated by counting links in the trophic (food) web," 39.

31. Ibid. Even if in "the real world," complexity was usually associated with greater stability, May argued, this did not mean that complexity begets stability, "no causal arrow need point from complexity to stability." To the contrary, he argued, "if there is a generalization it could be that [environmental] stability permits complexity," 76.

32. Ibid., 77.

33. See, for example, S. J. McNaughton, "Diversity and Stability of Ecological Communities: A Comment on the Role of Empiricism in Ecology," *American Naturalist* 111 (1977):515–25. Pimm argues that the differences between theoreticians (modellers) and field ecologists were not necessarily incompatible. Pimm, *Balance of Nature?*, 13.

34. See Connell and Sousa, "On the Evidence Needed to Judge Ecological Stability or Persistence," 789–824.

35. See Joseph Connell, "Diversity in Tropical Rain Forests and Coral Reefs," *Science* 199 (1978):1302–10.

36. Ibid., 1304.

37. See John F. Fox, "Intermediate Disturbance Hypothesis," *Science* 204 (1979):1344–45.

38. Connell, "Diversity in Tropical Rain Forests and Coral Reefs," 1308.

39. As Connell concluded in 1978: "My argument is that the assemblage of those organisms which determine the basic physical structure of two tropical communities (rain forest trees and corals) conform more closely to the non-equilibrium model. For these organisms, resource requirements are very general: inorganic substances (water, carbon dioxide, minerals) plus light and space, and, for corals, some zooplankton. It is highly unlikely that these can be partitioned finely enough to allow 100 or more species of trees to be packed, at equilibrium, on a single hectare. Instead, if competition is allowed to proceed unchecked, a few species eliminate the rest." Ibid., 1308–09.

40. See Jeremy Jackson, "Interspecific Competition and Species' Distributions: The Ghost of Theories and Data Past," *American Zoologist* 21 (1981):889–901. Idem, "Community Unity?" *Science* 264 (1994):1412–13; idem, "Does Ecology Matter?" *Paleobiology* 14 (1988):307–12.

41. Connell, "Diversity in Tropical Rain Forests and Coral Reefs,"1307.

42. Ibid.

43. May, *Stability and Complexity in Model Ecosystems*, iv.

44. Roger Bradbury, "Lessons of Crown-of-thorns Research," *Search* 7 (1976):461–62; 461.

45. Ibid., 462.

46. See A. M. Cameron, "*Acanthaster* and Coral Reefs: Population Outbreaks of a Rare and Specialized Carnivore in a Complex High-Diversity System," *Proceedings of the 3rd International Coral Reef Symposium* 1 (1977):193–99.

47. D. C. Potts, "Crown-of-thorns Starfish—Man-induced Pest or Natural Phenomenon?" in R. L. Kitching and R .E. Jones, eds., *The Ecology of Pests: Some Australian Case Histories* (Melbourne: CSIRO, 1981), 55–86.

48. Donald Potts, interview, Panamá, June 25, 1996.

49. Potts' work on coral was a follow-up from his work on land snails. His question was: On what sort of scales does natural selection occur in the field? For the snails, he had evidence that natural selection operating locally could explain much of the ecological variation he observed in populations in close proximity. But snails "ran around too much." Corals stayed in one place. Moreover, he could carry out experiments on reciprocal transplants of cloned organisms. He visited the Heron Island station, collected corals in one habitat, split them up into different pieces, transplanted each of the pieces in different habitats to determine whether selection was operating differently in different habitats.

50. Potts, "Crown-of-thorns Starfish—Man-induced Pest or Natural Phenomenon?," 57.

51. Ibid., 56

52. Ibid., 80.

53. Ibid., 65–66.

54. Ibid., 80.

55. Ibid., 81.

56. Ibid.

57. Peter Glynn, "Some Physical and Biological Determinants of Coral Community Structure in the Eastern Pacific," *Ecological Monographs* 46 (1976):431–56; idem, "Interactions between *Acanthaster* and *Hymenocera* in the Field and Laboratory," *Proceedings of the 3rd International Coral Reef Symposium* 1 (1979):209–15.

58. R. F. G. Ormond, A. C. Campbell, S. H. Head, R. J. Rainbow, and A. P. Saunders, "Formation and Breakdown of Aggregations of the Crown-of-thorns Starfish, *Acanthaster planci* (L.)," *Nature* 246 (1973): 167–69.

59. Potts, "Crown-of-thorns starfish—Man-induced Pest or Natural Phenomenon?," 76.

60. Ibid., 73.

61. Ibid.

62. Ibid.

63. Ibid., 73–74.

64. Ibid., 74.

65. Ibid., 75.

66. Ibid., 77.

67. Ibid., 81.

68. Ibid.

69. Ibid.

70. Ibid.

71. Charles Birkeland and John Lucas, *Acanthaster planci: Major Management Problem of Coral Reefs* (Boca Raton: CRC Press, 1990).

72. R. T. Paine, "Food Web Complexity and Species Diversity," *American Naturalist* 100 (1966):65–75; idem, "A Note on Trophic Complexity and Community Stability."

73. Charles Birkeland, "Interactions between a Sea Pen and Seven of its Predators," *Ecological Monogaphs* 44 (1974):211–32.

74. Charles Birkeland, Interview, Panamá, June 24, 1996.

75. Charles Birkeland, letter to the author, April 27, 1995.

76. Ibid.

77. Birkeland, interview.

78. Charles Birkeland, e-mail to the author, May 28, 1996.

79. Charles Birkeland, "Terrestrial Runoff as the Cause of *Acanthaster* Outbreaks," *American Zoologist* 20 (1980):811; idem, "Terrestrial Runoff as a Cause of Outbreaks of *Acanthaster planci* (*Echinodermate: Asteroidae*)," *Marine Biology* 69 (1982):175–85; 175.

80. Birkeland, "Terrestrial Runoff As a Cause of Outbreaks," 177.

81. Charles Birkeland, "*Acanthaster* in the Cultures of High Islands," *Atoll*

Research Bulletin No. 255 (1981):55–58; 55.

82. Ibid.

83. John Lucas e-mail to the author, February 19, 1997.

84. J. S. Lucas, "Quantitative Studies of Feeding and Nutrition During Larval Development of the Coral Reef Asteroid *Acanthaster planci* (L.)," *Journal of Experimental Marine Biology* 65 (1982):173–93.

85. Birkeland, "Terrestrial Runoff as a Cause of Outbreaks," 183.

86. T. C. R. White, "Weather, Food and Plagues of Insects," *Oecologia* 22 (1976):119–34; H. Wolda, "Seasonal Fluctuations in Rainfall, Food and Abundance of Tropical Insects," *Journal of Animal Ecology* 47 (1978):369–81. Idem, "Fluctuations in Abundance of Tropical Insects," *American Naturalist* 112 (1978):1017–55.

87. Birkeland, "Terrestrial Runoff As a Cause of Outbreaks," 183.

88. Charles Birkeland, "Large-Scale Fluctuations of Tropical Marine Populations—Are They Natural Events?," *The Siren*, No. 22 (1983):13–17.

89. Ibid., 14.

90. Ibid., 15.

91. Ibid., 14.

92. Ibid., 17.

93. Ibid.

94. See, for example, Charles Birkeland, "The Faustian Traits of the Crown-of-Thorns Starfish," *American Scientist* 77 (1989):155–63.

95. Birkeland and Lucas, *Acanthaster planci: Major Management Problem of Coral Reefs*, 11.

96. By the late 1980s, the crown-of-thorns problem there became overshadowed by public concern over the construction of the Ishigaki Airport. Ibid.

Chapter 11

1. Richard Kenchington, "The Crown-of-thorns Crisis in Australia: A Retrospective Analysis," *Environmental Conservation* 5 (1978):11–20; 18.

2. Gregg Borschmann, "Reef Killer Plague," *Sunday Sun*, 16 October 1983, 1, 48–49.

3. Gregg Borschmann, "Are We Heading for The Great Barren Reef," *The Age*, 17 October 1983.

4. Jeff Sampson, "Starfish Plague Threatens Reef," *Cairns Post*, 3 January 1984.

5. Borschmann, "Are We Heading for The Great Barren Reef."

6. Theo W. Brown, "Threat to the Reef," *The Australian*, 14 November 1983.

7. Theo Brown, "Time to Tackle 'Thorny' Issue," *Weekend Australian*, 17/18 December 1983.

8. Ibid.

9. Ibid.

10. John Lucas, "Scientists at Sea over Starfish," *The Australian*, 29 November 1983.

11. News release, "Crown-of-Thorns on the Great Barrier Reef," Minister for Home Affairs and Environment, the Hon. Barry Cohen, 19 January 1984, 1–3.

12. Crown-of-thorns Advisory Committee Report, 1985, S.6, 1

13. Ibid., 2.

14. Ibid., 3.

15. Ibid., 1.

16. Graeme Kelleher, "Crown of Thorns Starfish," Media Release, Great Barrier Reef Marine Park Authority, February 4, 1984, 1–6; 3. See also R. A. Kenchington and R. Pearson, "Crown of thorns Starfish on the Great Barrier Reef: A Situation Report," *Proceedings of the 4th International Coral Reef Symposium* 2 (1981):597–600; W. Nash and L. D. Zell, "*Acanthaster* on the Great Barrier Reef: Distribution of Five Transects Between 14 S and 18 S"; *Proceedings of the Fourth International Coral Reef Symposium* 2 (1981): 601–605.

17. Kelleher, "Crown of Thorns Starfish," 4.

18. Ibid., 5.

19. Andrew Stewart, "March of the Starfish," *The Sun-Herald*, 5 February 1984.

20. Editorial, "Starfish Plague," *Cairns Post*, 4 January 1984.

21. Ibid.

22. Anon., "'No Panic' on Crown of Thorns—Reef Authority," *Gladstone Observer*, 6 February 1984, 4.

23. Anon., "Crown-of-thorns—No Threat to Reef," *Daily Mercury*, 6 February 1984; Anon., "Crown of Thorns Problem Not as Bad as It Seems," *Townsville Daily Bulletin*, 6 February 1984; "Science Gives Life Back to the Reef," *Daily Sun*, 11 February 1984; James Guthrie, "Starfish Claims Rejected," *Cairns Post*, 6 February 1984.

24. Anon., "'Whitewash' Alleged on Starfish Reef Threat," *Townsville Daily Bulletin*, 9 February 1984; Anon., "Starfish Threat a 'Whitewash,'" *Courier Mail*, 9 February 1984.

25. W. Nash and L. D. Zell, "*Acanthaster* on the Great Barrier Reef: Distribution of Five Transects Between 14 S and 18 S," 605.

26. John Austin interview with Dr. Anne Cameron, "The Crown of Thorns Starfish," *A.M.*, 10 February 1984, Transcript of tape recording by Department of the Parliamentary Library, Parliament of the Commonwealth of Australia. The next week a skindiver Betty Young appeared on the same radio program to describe dead reefs, and to rebuke the Great Barrier Reef Marine Park Authority "for taking no action." David Morgan, interview with Betty Young, skindiver, "Great Barrier Reef: More Evidence that the Crown of Thorns Starfish is a Threat, Disputed by the Marine Park Authority," *A.M.*, 18 February 1984. Transcript of tape recording by Department of the

Parliamentary Library, Parliament of the Commonwealth of Australia. Obtained from Charles Birkeland.

27. Robert Raymond, "A Conspiracy of Silence over the Destruction of the Reef," *The Bulletin*, December 18, 1984, 68–77.

28. Ibid., 68.

29. Ibid., 71.

30. Ibid.

31. Ibid., 69.

32. Ibid., 76.

33. Ibid.

34. Ibid., 77. John Lucas, "The Biology, Exploitation, and Mariculture of Giant Claims (*Tridacnidae*)," *Reviews in Fisheries Science* 2 (1994):181–223. By 1983, eight species of giant clams had been overfished for meat and shells. The two largest species were seriously threatened due in part to international poaching for their abductor muscle. Stimulated by declining giant claim stocks, local extinctions, and its commercial value, mariculture methods for giant claim farming were explored. Conservation techniques such as restocking reefs with cultured claims was also considered to be the most direct method of enhancing populations in the wild. Others, including Richard Chesher working in Tonga, promoted the idea of placing sparse adult giant clams in aggregates with the hope of facilitating their reproductive success. For a discussion of the conflict between the approaches of Chesher and Lucas, see Tim Cahill, "The Clam Scam," *Outside* June (1991):48–53, 108–15.

35. Raymond, "A Conspiracy of Silence," 77.

36. Ibid.

37. Robert Raymond, *Starfish Wars: Coral Death and the Crown-of-Thorns* (Melbourne: Macmillan, 1986).

38. R. H. Bradbury, T. J. Done, S. A. English, D. A. Fisk, P. J. Moran, R. E. Reichelt, and D. Mcb. Williams, "The Crown-of-Thorns: Coping with Uncertainty," *Search* 16 (1985):106–09.

39. P. J. Moran, R. H. Bradbury and R. E. Reichelt, "Distribution of Recent Outbreaks of the Crown-of-Thorns Starfish (*Acanthaster planci*) along the Great Barrier Reef: 1985–1986," *Coral Reefs* 7 (1988):125–37.

40. Ibid., 131.

41. Roger Bradbury, interview, Panamá, June 25, 1996.

42. Ibid.

43. "Report of the Crown of Thorns Advisory Committee, 11 January 1985" (K. J. Back chairman), Townsville. Available from Great Barrier Reef Marine Park Authority. Obtained from Charles Birkeland.

44. Richard Kenchington, "*Acanthaster planci* and Management of the Great Barrier Reef," *Bulletin of Marine Science* 4 (1987):552–60.

45. Ibid., 552–53. See R. Endean and A. M. Cameron, "Ecocatastrophe on the Great Barrier Reef," *Proceedings of the 5th International Coral Reef*

Congress 5 (1985):309–14.

46. See R. G. Pearson, "Recovery and Recolonization of Coral Reefs," *Marine Ecology Progress* Series 4 (1981):105–22; M. W. Colgan, "Succession and Recovery of a Coral Reef after Predation by *Acanthaster planci* (L.)," *Proceedings of the 4th International Coral Reef Symposium* 2 (1982):33–38.

47. Kenchington, "*Acanthaster planci* and Management," 556.

48. V. J. Harriot and P. L. Harrison, "Methods to Accelerate the Recolonization of Corals in Damaged Reef Systems," Report to the Great Barrier Reef Marine Park Authority, Townsville, 1984.

49. P. D. Cheney, "An Analysis of the *Acanthaster* Control Programs in Guam and the Trust Territories of the Pacific Islands," *Micronesica* 9 (1973):171–80.

50. Kenchington, "*Acanthaster planci* and Management," 555.

51. See R. H. Bradbury, L. S. Hammond, P. J. Moran, and R. E. Reichelt, "The Stable Points, Stable Cycles and Chaos of the *Acanthaster* Phenomenon: A Fugue in Three Voices," *Proceedings of the 5th International Coral Reef Symposium* 5 (1985):303–08.

52. See Charles Birkeland and John S. Lucas, *Acanthaster planci: Major Management Problem of Coral Reefs* (Boca Raton, CRC Press, 1990), 197.

53. M. Yamaguchi, "*Acanthaster planci* infestations of reefs and coral assemblages in Japan: A Retrospective Analysis of Control Efforts," *Coral Reefs* 5 (1986):277–88.

54. Ibid., 227.

55. Birkeland and Lucas, *Acanthaster planci: Major Management Problem of Coral Reefs*, 197. For a discussion of the circumstances in which control programs are likely to be feasible, see R. E. Reichelt, "Dispersal and Control Models of *Acanthaster planci* Populations on the Great Barrier Reef," in R. E. Bradbury ed., *Acanthaster and the Coral Reef: A Theoretical Perspective* (Berlin: Springer-Verlag 1990), 6–16.

56. Kenchington, "*Acanthaster planci* and Management," 559.

57. Leon Zann and Elaine Eager, eds., "The Crown of Thorns Starfish," *Australian Science Magazine* 3 (1987):15–54; 17. AIMS also issued similar booklets. See for example, Peter Moran, *Crown-of-thorns Starfish Questions and Answers* (The Australian Institute for Marine Science, 1988).

58. D. T. Anderson, "Crown of thorns starfish. A Report to the Minister for the Arts, Sport, the Environment, Tourism and Territories, Senator the Honourable Graham Richardson," University of Sydney, 1989. Cited in Birkeland and Lucas, *Acanthaster planci: Major Management Problem of Coral Reefs*, 11.

Chapter 12

1. J. Cherfas, "Ecology Invades a New Environment," *New Scientist* 1583 (1987):42-46.

2. Ibid., 43.

3. Ibid., 44.

4. Ibid.

5. Ibid. See also Stuart Pimm, *The Balance of Nature? Ecological Issues in the Conservation of Species and Communities* (Chicago: University of Chicago Press, 1991), 13–14.

6. Cherfas, "Ecology Invades a New Environment," 45.

7. Ibid.

8. Ibid.

9. Ibid., 46.

10. Ibid.

11. Pimm, *The Balance of Nature?*, xi.

12. P. R. Ehrlich and A. Ehrlich, *Extinction: the Causes and Consequences of the Disappearance of Species* (New York: Random House, 1981).

13. Pimm, *Balance of Nature?*, xi.

14. Ibid., 17.

15. Ibid., 16.

16. E. Frankel, "Evidence from the Great Barrier Reef of Ancient *Acanthaster* Aggregations," *Atoll Research Bulletin* 220 (1987):75–86; 84.

17. Charles Birkeland and John S. Lucas, *Acanthaster Planci: Major Management Problem of Coral Reefs* (Boca Raton: CRC Press, 1990), 202.

18. Charles Birkeland, "The Faustian Traits of the Crown-of-Thorns Starfish," *American Scientist* 77 (1989):155–63.

19. Ibid., 162.

20. P. J. Moran and R. H. Bradbury, "The Crown-of-Thorns Starfish Controversy," *Search* 20 (1989):5.

21. Ibid., 6.

22. Ibid.

23. Ibid. See also P. J. Moran, "The *Acanthaster* Phenomenon," *Oceanography and Marine Biology: An Annual Review* 24 (1986):379–480; 458.

24. P. J. Moran, R. E. Reichelt, and R. H. Bradbury, "An Assessment of the Geological Evidence for Previous Acanthaster Outbreaks," *Coral Reefs* 4 (1986):235–38; 235. Frankel's "reality check" was that he could detect outbreaks on reefs where scientists knew they occurred in the 1960s. However, their statistical reassessment of Frankel's evidence showed that the occurrence of *Acanthaster* remains in sediments was independent of whether or not the reef from which the sample was collected had experienced a recent outbreak.

25. P. D. Walbran, R. A. Henderson, A. J. T. Jull, and M. J. Head, "Evidence from Sediments of Long-term *Acanthaster planci* Predation on Corals of the Great Barrier Reef," *Science* 245 (1989): 847–50.

26. Ibid., 850.

27. J. K. Keesing, R. H. Bradbury, L. M. De Vatier, M. J. Riddle and G. De'ath, "Geological Evidence for Recurring Outbreaks of the Crown-of-thorns Starfish: A Reassessment from an Ecological Perspective," *Coral Reefs* 11 (1992):79–85.

28. Moran, "The *Acanthaster* Phenomenon," 458.

29. A. M. Cameron and R. Endean, "Renewed Population Outbreaks of a Rare and Specialized Carnivore (the starfish *Acanthaster planci*) in a Complex High-diversity System," *Proceedings of the Fourth International Coral Reef Symposium* 2 (1982):593–96.

30. Ibid.

31. Charles Birkeland, "Terrestrial Runoff as a Cause of Outbreaks of *Acanthaster planci* (*Echinodermata: Asteroidea*)," *Marine Biology* 69 (1982): 175–85.

32. For example, Dana, Newman, and Fager maintained that large starfish populations were first observed north of Townsville in 1959, thus fitting the weather conditions they prescribed. Yet the first reported outbreaks did not occur until 1962—on Green Island. See Thomas F. Dana, William A. Newman, and Edward W. Fager, "*Acanthaster* Aggregations: Interpreted as Primarily Responses to Natural Phenomena," *Pacific Science* 26 (1972): 355–72; 368.

33. Moran, "The *Acanthaster* Phenomenon," 461. See also Birkeland and Lucas, "*Acanthaster planci*: Major Management Problem of Coral Reefs," 199.

34. In 1982, Birkeland added a new important criticism—the mechanism proposed by Endean would lead to a gradual increase in starfish numbers over several years, whereas observations in the field indicated that outbreaks build up very suddenly. See Birkeland, "Terrestrial Runoff as a Cause of Outbreaks."

35. R. Endean, "Crown-of-thorns Starfish on the Great Barrier Reef," *Endeavour* 6 (1982):10–14.

36. Moran, "The *Acanthaster* Phenomenon," 470.

37. Ibid.

38. Roger Bradbury, interview, Panamá, June 26, 1996. See R. H. Bradbury, L. S. Hammon, P. J. Moran and R. E. Reichelt, "Coral Reef Communities and Crown-of-thorns Starfish: Evidence for Qualitative Stable Cycles," *Journal of Theoretical Biology* 113 (1985):69–80.

39. R. F. G. Ormond, A. C. Campbell, S. H. Head, R. J. Moore, P. R. Rainbow, and A. P. Saunders, "Formation and Breakdown of Aggregations of the Crown-of-thorns Starfish, *Acanthaster planci* (L.)," *Nature* 240 (1973):167–69; 168.

40. See Rupert Ormond, Roger Bradbury, Scott Bainbridge, Katarina Fabricius, John Keesing, Lyon de Vantier, Paul Medlay, and Andrew Steven, "Test of a Model of Regulation of Crown-of-Thorns Starfish by Fish Predation," in R. H. Bradbury, ed., *Acanthaster and the Coral Reef: A Theoretical Perspective. Lecture Notes in Biomathematics* 88 (Berlin: Springer-Verlag, 1990), 189–207; 191.

41. Ibid.

42. Ibid., 198.

43. Ibid. See also, A. D. Steven, "An Analysis of Fishing Activities on Possible Predators of Crown-of-thorns starfish *Acanthaster planci* on the Great Barrier Reef," Report to the Great Barrier Reef Marine Park Authority, Townsville, Queensland, 1988.

44. P. L. Antonelli and N. D. Kazarinoff, "Starfish Predation of a Growing Coral Reef Community," *Journal of Theoretical Biology* 107 (1984):667–84.

45. Bradbury, ed., *Acanthaster and the Coral Reef.*

46. Ibid., iii.

47. Rupert Ormond, Roger Bradbury, Scott Bainbridge, Katarina Fabricius, John Keesing, Lyon de vantier, Paul Medlay, and Andrew Steven, "Test of a Model of Regulation of Crown-of-Thorns Starfish by Fish Predation," in Bradbury, *Acanthaster and the Coral Reef,* 189–207.

48. Ibid., 202.

49. Ibid., 203.

50. R. H. Bradbury and P. L. Antonelli, "What Controls Outbreaks?" in *Acanthaster and the Coral Reef. A Theoretical Perspective,* 278–90. As developed by René Thom in the mid-1970s catastrophe theory describes the behaviour of systems where "control variables" interact with "state variables" such as food to produce abrupt and discontinuous changes—the catastrophes. See René Thom, *Structural Stability and Morphogenesis* (New York: Benjamin Addison-Wesley, 1975).

51. Bradbury and Antonelli, "What Controls Outbreaks?," 288.

52. Roger Bradbury, "Understanding *Acanthaster,*" *Coenoses* 6 (1991): 121–26; 123.

53. Russell Reichelt, "Dispersal and Control Models of *Acanthaster Planci* Populations on the Great Barrier Reef," in Bradbury, ed., *Acanthaster and the Coral Reef: A Theoretical Perspective,* 6–16.

54. T. J. Done, "Transition Matrix Models, Crown-of-thorns and Corals," 291–97. See also T. J. Done, "Simulation of the Effects of *Acanthaster planci* on the Population Structure of Massive Coral in the Genus *Porites*: Evidence of Population Resilience?" *Coral Reefs* 6 (1987):75–90; T. J. Done, "Simulation of Recovery of Pre-Disturbance Size Structure in Populations of *Porites*. Damage by the Crown-of-thorns Starfish *Acanthaster planci.*" *Marine Biology* 100 (1988):51–61; T. J. Done, K. Osborne, K. F. Navin "Recovery of Corals Post-Acanthaster Progress and Prospects," *Proceedings of the 6th International Coral Reef Symposium* 2 (1988):137–42.

55. Birkeland and Lucas, *Acanthaster planci: Major Management Problem of Coral Reefs,* 202.

56. K. H. Mann, *Ecology of Coastal Waters: A Systems Approach,* Studies in Ecology. Vol. 8. (Berkeley: University of California Press, 1982). "The approach used in this book differs from that used in many others, in concentrating first and foremost on ecological processes rather than on organisms or populations. In fact, when experimenting with earlier versions, I wondered whether it would be possible to make a useful contribution without naming a

plant or an animal," ix.

57. Birkeland, "The Faustian Traits of the Crown-of-Thorns Starfish," 157.

58. Birkeland and Lucas, *Acanthaster planci: Major Management Problem of Coral Reefs*, 2.

59. Birkeland, "The Faustian Traits of the Crown-of-Thorns Starfish," 155.

60. Ibid.

61. Birkeland and Lucas, *Acanthaster planci: Major Management Problem of Coral Reefs*, 202.

62. Bradbury, "Understanding *Acanthaster*," 123, 124.

63. Birkeland and Lucas, *Acanthaster planci: Major Management Problem of Coral Reefs*, 199.

64. Ibid., 125.

65. Y. I Sorokin, *Coral Reef Ecology. Ecological Studies 102* (Berlin: Springer-Verlag, 1993), 400.

Chapter 13

1. Peter Hulm and John Pernetta, eds., *Reefs at Risk: A Programme of Action.* October, 1993. This booklet was produced by the Marine and Coastal Areas Programme of ACNE—The World Conservation Union with financial assistance from WWF—The World Wide Fund for Nature (formerly The World Wildlife Fund) as a contribution to the Long-term Global Monitoring System of Coastal and Near-Shore Phenomena related to Climate Change, sponsored by the United Nations Environment Program (UNEP), the Intergovernmental Oceanographic Commission (IOC) of UNESCO, the World Meterological Organization (WMO) and ACNE—The WORLD Conservation Union. It is based on the Report of the Global Task Team on the Implications of Climate Change on Coral Reefs. The text was prepared and edited by Peter Hulm and John Pernetta with contributions from R. W. Buddemeier and Clive Wilkinson, co-editors of the Task Team report.

2. Ibid.

3. Ibid., 9. See also Sue Wells, "A Future for Coral Reefs," *New Scientist*, October 30, (1986):46–50; C. S. Rogers, "Degradation of Caribbean and Western Atlantic Coral Reefs and Decline of Associated Reef Fisheries," *Proceedings of the 5th International Coral Reef Symposium*, Papeete, Tahiti 6 (1985):491–96; C. R. Wilkinson, "Coral Reefs Are Facing Widespread Extinctions: Can We Prevent these through Sustainable Management Practices?" *Proceedings of the Seventh International Coral Reef Symposium*, Guam, June 1992, 1 (1994):11–21.

4. Y. I Sorokin, *Coral Reef Ecology. Ecological Studies 102* (Berlin: Springer-Verlag, 1993), 415.

5. International Coral Reef Initiative (ICRI), *Report to the United Nations Commission on Sustainable Development*, 1996, 2.

6. Hulm and Pernetta, *Reefs at Risk*, 20.

7. Peter Glynn, "Coral Reef Bleaching: Facts, Hypotheses and Implications," *Global Change Biology*, 2 (1996):495–509.

8. Barbara E. Brown and John C. Ogden, "Coral Bleaching," *Scientific American* 268 (1993):64–70; 69; Richard Grigg, "Coral Reef Environmental Science: Truth Versus the Cassandra Syndrome," *Coral Reefs* 11 (1992):183–86; 184; see also J. E. N. Vernon, *Corals of Australia and the Indo-Pacific* (North Ryde NSW: Angus and Robertson, 1986), 45–61.

9. Bleaching patterns may also reflect varying sensitivities of diverse varieties of zooxanthella within and among host species at different depths. See R. Rowan and N. Knowlton, "Intraspecific Diversity and Ecological Zonation in Coral-Algal Symbiosis," *Proceedings of the National Academy of Sciences USA* 92 (1995):2850–53. There may be hundreds of species of symbiotic algae living in coral. See also R. Rowan and D. A. Powers, "A Molecular Genetic Classification of Zooxanthellae and the Evolution of Animal-Algal Symbioses," *Science* 251 (1991):1348–51.

10. The mechanism of algal release was not fully understood. Nor was the reason for the release. One hypothesis held that stressed coral polyps provide fewer nutrients (because of lowered metabolism) to the algae—as a result, the algae abandon their place of residence. Another suggested that algae emit poisonous substances when they experience adverse conditions and that these toxins may deleteriously affect the host. See O. Hoegh-Guldberg, L. R. McCloskey, and L. Muscatine, "Expulsion of Zooxanthellae by Symbiotic Enidarians from the Red Sea," *Coral Reefs* 5 (1987):201–04.

11. Hulm and Pernetta, *Reefs at Risk*, 7.

12. See Rowan and Knowlton, "Intraspecific diversity and Ecological Zonation in Coral-algal Symbiosis," 2850–53. See also Rowan and Powers, "A Molecular Genetic Classification of Zooxanthellae and the Evolution of Animal-Algal Symbioses," 1348–51.

13. Thomas Goreau, "Mass Expulsion of Zooxanthellae from Jamaican Reef Communities after Hurricane Flora," *Science* 145 (1964):383–86.

14. See Peter Glynn, "Coral Reef Bleaching: Ecological Perspectives," *Coral Reefs* 12 (1993):1–17; 2.

15. Brown and Ogden, "Coral Bleaching."

16. N. E. Graham and W. B. White, " The El Niño Cycle: A Natural Oscillator of the Pacific Ocean—Atmosphere System," *Science* 240 (1988): 1293–1302.

17. D. Halpern, S. P. Hayes, A. Leetmaa, D. V. Hansen, and S. G. H Philander, "Oceanographic Observations of the 1982 Warming of the Tropical Eastern Pacific," *Science* 221 (1983):1173–75; R. A. Kerr, "Fading El Niño Broadening Scientist's View," *Science* 221 (1983):940–41.

18. P. W. Glynn, "Extensive 'Bleaching' and Death of Reef Corals on the Pacific Coast of Panama," *Environmental Conservation* 10 (1983):149–54.

19. Glynn also noted that the reefs affected by bleaching off the Galápagos

were subsequently eroding at an estimated rate of 2.5 to 5 centimetres a year, due to invasions of sea urchins that ground the coral limestone down to sand as they searched for food. See Leslie Roberts, "Coral Bleaching Threatens Atlantic Reefs," *Science* 238 (1987):1228–29; Peter W. Glynn, "Coral Reef Bleaching in the 1980s and Possible Connections with Global Warming," *Trends in Ecology and Evolution* 6 (1991):175–78.

20. In laboratory tests, Glynn and Luis D'Croz of the University of Panamá showed that bleaching of the major reef-building coral in the eastern Pacific, *Pocillopora damicornis*, was induced at 30–32°C, the same temperature it did in the field. P. W. Glynn and L. D'Croz, "Experimental Evidence for High Temperature Stress as the Cause of El Niño—coincident Coral Mortality," *Coral Reefs* 8 (1990):181–91.

21. B. E. Brown and K. W. Suharsono, "Damage and Recovery of Coral Reefs Affected By El Niño Related Seawater Warming in the Thousand Islands, Indonesia," *Coral Reefs* 8 (1990):163–70.

22. E. H. Williams, Jr., C. Goenaga, and V. Vicente, "Mass Bleaching on Atlantic Coral Reefs," *Science* 238 (1987):877–78. E. H. Williams, Jr. and L. Bunkley-Williams, "Bleaching of Caribbean Coral Reef Symbionts in 1987–1988," *Proceedings of the Sixth International Coral Reef Symposium* 3 (1988):313–18. There were also predictions of future mass bleaching for the reefs surrounding the Hawaiian islands in 1989. See P. L. Jokeil and S. L. Coles, "Response of Hawaiian and other Indo-Pacific Reef Corals to Elevated Temperature," *Coral Reefs* 8 (1990):155–62.

23. In the bleaching events of 1982–83 and 1987, reefs in Belize and many neighboring nations in the western Caribbean/Gulf of Mexico region had not been affected. However, on November 9, 1995, NOAA issued a release that mass bleaching had occurred on the western hemisphere's longest and most pristine barrier reef in Belize as well as other areas of the western Caribbean and Gulf of Mexico. See Constance Holden, "Reef Bleaching Spreads in Caribbean," *Science* 270 (1995):919.

24. Leslie Roberts, "Coral Bleaching Threatens Atlantic Reefs," *Science* 238 (1987):1228–29.

25. Stephen H. Schneider, "The Global Warming Debate: Science or Politics," *Environ. Sci. Technol.* 24 (1990):432–35; 434.

26. See Schneider, "The Global Warming Debate"; Jeremy Leggett, ed., *Global Warming: The Green Peace Report* (Oxford: Oxford University Press, 1990), 2. There is agreement today that Earth's surface temperature has increased over the last 100 years by between 0.3° and 0.6°C. See J. T. Houghton et al., eds., *Climate Change 1995—The Science of Climate Change* (Cambridge: Cambridge University Press, 1995). Mark A. Cane et al., "Twentieth-Century Sea Surface Temperature Trends," *Science* 275 (1997): 957–60.

27. This was the consensus on climate change used for the workshop discussion on coral bleaching and climate change in Miami in 1990. See Chris-

topher F. D'Elia, Robert W. Buddemeier, and Stephen V. Smith, eds., "Workshop on Coral Bleaching, Coral Reef Ecosystems and Global Change: Report of Proceedings," A Maryland Sea Grant College Publication Um-SG-TS-91-03, 1991, 29.

28. Hulm and Pernetta, *Reefs at Risk*, 15.

29. Ibid. See also Sue Wells and Alasdair Edwards, "Gone with the Waves," *New Scientist*, November 11 (1989):47–81; John Connell and John Lea, "My Country Will not be There," *Cities*, November (1982): 295–309.

30. For some strong examples of such views, see Warren T. Brookes, "The Global Warming Panic," *Forbes*, December 25 (1989):96–102; Ben Bolch and Harold Lyons, *Apocalypse Not: Science, Economics and Environmentalism* (Washington, D.C.: CATO Institute, 1993).

31. P. D. Jones, and T. M. L. Wigley, "Global Warming Trends," *Scientific American*, August (1990):84–91; E. Friis-Christensen and K. Lassen, "Length of the Solar Cycle: An Indicator of Solar Activity Closely Associated with Climate," *Science* 254 (1991): 698–700; G. C. Reid, "Solar Total Irradiance Variations and the Global Sea Surface Temperature Record," *Journal of Geophysical Research* 96 (1991): 2835–44.

32. Robert Jastrow, William Nierenberg, and Frederick Seitz, *Scientific Perspectives on the Greenhouse Problem* (Ottawa, Ill.: The Marshall Press, 1990); Aaron Wildavsky, *But Is It True? A Citizen's Guide to Environmental Health and Safety Issues* (Cambridge: Harvard University Press, 1995), 340–73.

33. Philip Anderson, "Uncertainties about Global Warming," *Science* 247 (1990):1529.

34. Brown and Ogden, "Coral Bleaching," 64.

35. See E.H. Williams Jr. and Lucy Bunkley-Williams, "The World-Wide Coral Reef Bleaching Cycle and Related Sources of Coral Mortality," *Atoll Research Bulletin* 335 (1990):1–71; 1.

36. Barbara E. Brown, ed., "Coral Bleaching." *Special Issue of Coral Reefs* 8 (1990).

37. E. F. Hollings, (Chair) 1988. "Bleaching of Coral Reefs in the Caribbean. Oral and Written Testimony to the Commerce, Justice, States, Judiciary, and Related Agencies, Appropriations Subcommittee, U.S.A. Senate," 10 November, 1987. Cited in Williams and Bunkley-Williams, "The World-Wide Coral Reef Bleaching Cycle," 1; Leslie Roberts, "Warm Waters, Bleached Corals," *Science* 241 (1990):12.

38. Christopher F. D'Elia, Robert W. Buddemeier, and Stephen V. Smith, "Workshop on Coral Bleaching, Coral Reef Ecosystems and Global Change: Report of Proceedings," A Maryland Sea Grant College Publication Um-SG-TS-91-03, 1991, 3–4.

39. See Peter W. Glynn, "Coral Reef Bleaching in the 1980s and Possible Connections with Global Warming," *Trends in Ecology and Evolution* 6 (1991): 175–78; 176. By 1996, Glynn argued that present evidence indi-

cated that large-scale coral beaching was largely due to elevated sea temperatures and high solar irradiance (especially ultraviolet wavelengths). Glynn,
"Coral Reef Bleaching: Facts, Hypotheses, and Implications."

40. Hulm and Pernetta, eds., *Reefs at Risk*, 21. See also Leslie Roberts,
"Corals Remain Baffling," *Science* 239 (1988):256.

41. Roberts, "Warm Waters, Bleached Corals."

42. Thomas J. Goreau, "Coral Bleaching in Jamaica," *Nature* 343
(1990):417. See also Thomas J. Goreau, Raymond Hayes, Jenifer Clark,
Daniel J. Basta, and Craig Robertson, "Elevated Sea Surface Temperatures
Correlate with Caribbean Coral Bleaching," in Richard A. Geyer, ed., *A
Global Warming Forum: Scientific, Economic, and Legal Overview* (Boca
Raton: CRC Press, 1993), 225–55.

43. P. J. Jokiel and S. L. Coles, "Responses of Hawaiian and other
Indo-Pacific Reef Corals to Elevated Temperature," *Coral Reefs* 8
(1990):155–62; 161.

44. Lucy Bunkley-Williams and Ernest H. Williams, Jr., "Global Assault
on Coral Reefs," *Natural History* 4 (1990): 47–54; 52.

45. Ibid., 54.

46. E. H. Williams, Jr. and Lucy Bunkley-Williams, "The World-Wide
Coral Reef Bleaching Cycle and Related Sources of Coral Mortality," *Atoll
Research Bulletin* 335 (1990) 1–71; 1.

47. Bunkley-Williams and Williams, "Global Assault on Coral Reefs," 54.

48. See D.W. Kinsey, "The Greenhouse Effect and Coral Reefs," *Pacific
Science* 46 (1992):375–76; 376. See also Stephanie Pain, "Coral Reefs Will
Thrive in the Greenhouse," *New Scientist*, March 3, 1990, p. 30.

49. D'Elia, Buddemeier, and Smith, "Workshop on Coral Bleaching,
Coral Reef Ecosystems and Global Change: Report of Proceedings," 4.

50. Ibid., 20.

51. Ibid., 21.

52. To develop greater spatial and temporal coverage, some examined re-
motely sensed SST databases. But these also resulted in conflicting conclusions, due to a variety of underlying asumptions and degree of accuracy of
measurements. This conflict was especially evident in regard to the 1987 Caribbean mass bleaching event. See D. K. Atwood, J. C. Hendee, A. Mendez
"An Assessment of Global Warming Stress on Caribbean Coral Reef Ecosystems," *Bulletin of Marine Science* 51 (1992):118–30. Goreau et al., "Elevated
Sea Surface Temperatures Correlate with Caribbean Coral Reef Bleaching."

53. D'Elia, Buddenmeier, and Smith, "Workshop on Coral Bleaching:
Coral Reef Ecosystems and Global Change," 44.

54. Ibid., 45.

55. Ibid., 21.

56. Ibid., 10.

57. Ibid., 9. How resilient were coral reefs to bleaching events? How long
would recovery take? How much bleaching and how frequently did it have to

occur before it caused local extinctions? The effects of repeated bleaching could be investigated using models that had been constructed to investigate the same questions in regard to damage caused by the crown-of-thorns on the Great Barrier Reef. See T. J. Done, "Simulation of the Effects of *Acanthaster planci* on the Population Structure of Massive Corals in the Genus Porites: Evidence of Population Resilience?" *Coral Reefs* 6 (1987): 75–90.

58. D'Elia, Buddemeier, and Smith, "Workshop on Coral Bleaching, Coral Reef Ecosystems and Global Change: Report of Proceedings," 8.

59. Ibid., 28.

60. Ibid., 10.

61. Ibid.

62. Ibid.

63. Constance Holden, ed., "Reef Bleaching Spreads in Caribbean," *Science* 270 (1995):919.

64. Brown and Ogden, "Coral Bleaching," 69.

65. Ibid., 70.

66. Peter W. Glynn, "Coral Reef Bleaching in the 1980s and Possible Connections with Global Warming," *Trends in Ecology and Evolution* 6 (1991):175–78.

67. See Richard Grigg, "Coral Reefs and Environmental Change—The Next 100 Years: A Synopsis and Abstracts of Papers Presented at a Symposium of the XVII Pacific Science Congress," *Pacific Science* 46 (1992):374–79.

68. John Ogden, interview, March 10, 1997, Florida Institute of Oceanography, St. Petersburg.

69. John C. Ogden, "Cooperative Coastal Ecology at Caribbean Marine Laboratories," *Oceanus* 30 (1987):9–15. See also Ogden et al., "Caribbean Coastal Marine Productivity (CARICOMP): A Research and Monitoring Network of Marine Laboratories, Parks and Reserves," *Proceedings of the Eighth International Coral Reef Symposium*, Panamá, June 1996, 641–46.

70. See Ogden et al, "Caribbean Coastal Marine Productivity (CARICOMP)."

71. H. A. Lessios, D. R. Robertson, J. D. Cubit "Spread of *Diadema* mass mortality through the Caribbean," *Science* 226 (1984):335–37.

72. See H. A. Lessios, "*Diadema antillarum* 10 years after Mass Mortality: Still Rare, Despite Help from a Competitor," *Proceedings of the Royal Society of London* 256 (1995): 331–337; Idem, "Mass Mortality of *Diadema antillarum* in the Caribbean: What Have We Learned?" *Annual Review of Ecological Systems* 19 (1988):371–93.

73. See W. B. Gladfelter, "White Band Disease in *Acropora palmata*: Implications for the Stucture and Growth of Shallow Reefs," *Bulletin of Marine Science* 32 (1982):639–43.

74. Ogden et al, "Caribbean Coastal Marine Productivity (CARICOMP)."

75. PACICOMP has not gotten under way for lack of funding. Charles Birkeland, personal communication.

76. Ogden et al.

77. Ibid.

78. Ibid.

79. International Coral Reef Initiative, "Report to the United Nations Commission on Sustainable Development" 1996, 3.

82. Hulm and Pernetta, *Reefs at Risk*. See also International Coral Reef Initiative, "Report to the United Nations Commission of Sustainable Development."

83. International Coral Reef Initiative, "Report to the United Nations Commission on Sustainable Development," 11.

Chapter 14

1. Richard W. Grigg, "Coral Reef Environmental Science: Truth Versus the Cassandra Syndrome," *Coral Reefs* 11 (1992):183–86.

2. Ibid., 183.

3. Ibid.

4. Ibid.

5. L. Cornell and J. Surowiecki eds., *The Pulse of the Planet—A State of the Earth Report from the Smithsonian Institution Center for Short Lived Phenomena* (New York: Crown, 1972), 56.

6. Grigg, "Coral Reef Environmental Science," 183–84.

7. Grigg referred to P. D. Walbran, R. A. Henderson, J. W. Faithful, H. A. Polach, R. J. Sparks, G. Wallace, and D. C. Lowe, "Crown-of-thorns Starfish Outbreaks on the Great Barrier Reef: A Geological Perspective based on the Sediment Record," *Coral Reefs* 8 (1989):67–78.

8. See Daniel B. Blake, "The Affinities and Origins of the Crown-of-thorns Sea Star *Acanthaster* Gervais," *Journal of Natural History* 13 (1979): 303–14.

9. Grigg, "Coral Reef Environmental Science," 184.

10. Ibid., 185.

11. Ibid.

12. Robert White, "The Great Climate Debate," in Robert Jastrow, William Neirenberg, Frederick Seitz, *Scientific Perspectives on the Greenhouse Problem* (Ottawa, Ill.: The Marshall Press, 1990), 71–93, 71–72.

13. Laurie Garrett, *The Coming Plague. Newly Emerging Diseases in a World out of Balance.* (New York: Farrar, Straus and Giroux, 1994). See also Andrew Nikiforuk, *The Fourth Horseman. A Short History of Plagues, Scourges and Emerging Viruses* (London: Penguin, 1992); Robin Henig, *A Dancing Matrix. Voyages During the Viral Frontier* (New York: Knopf, 1993); Rodney Barker, *And the Waters Turned to Blood: The Ultimate Biological Threat* (New York: Simon and Schuster, 1997).

14. P. Boyer, "When the Time Shall Be No More: Prophecy Belief in

Modern American Culture," *Canadian Journal of Sociology* 19 (1994):273–76.

15. Michael Ludyanskiy et al., "Impact of the Zebra Mussel, a Bivalve Invader," *BioScience* 43 (1993):533–41; 533.

16. John H. Gibbon, "The Politics of Science," *Science* 269 (1995):143.

17. Robert Nealson, "Unoriginal Sin: The Judeo-Christian Roots of Ecotheology," *Policy Review*, Summer (1990):52–59.

18. Aaron Wildavsky, "Ecotheology Is Liberal, Not Religious," *Policy Review* Fall (1990): 90–91. See also B. Bolche and H. Lyons, *Apocalypse Not: Science, Economics and Environmentalism* (Washington, D.C.: CATO Institute, 1993), 6.

19. See B. Bolche and Harold Lyons, *Apocalypse Not: Science, Economics, and Envrionmentalism* (Washington, D.C.: CATO Institute, 1992). Even the concerns of plagues have also been dismissed as fear-mongering paranoia. See Malcom Gladwell, "The Plague Year," *The New Republic*, July 17 (1995):38–46.

20. Gregg Easterbrook, *A Moment on the Earth: The Coming Age of Environmental Optimism* (New York: Viking, 1995).

21. Garrett Hardin, *Living Within Limits. Ecology, Economics and Population Taboos* (New York: Oxford University Press, 1993).

22. Grigg, "Coral Reef Environmental Science," 185.

23. See Carl J. Walters and C. S. Holling, "Large-scale Management Experiments and Learning by Doing," *Ecology* 71 (1990):2060–68; 2066.

24. See W. C. Clark, D. D. Jones, and C. S. Holling, "Lessons for Ecological Policy Design: A Case Study of Ecosystem Management," *Ecological Modelling* 7 (1979):1–53; T. Royama, "Population Dynamics of the Spruce Budworm *Choristoneura Fumiferana*," *Ecological Monographs* 54 (1984): 429–62.

25. See Chris Elfring, "Yellowstone: Fire Storm over Fire Management," *BioScience* 39 (1989):667–72; Norman L. Christensen et al., "Interpreting the Yellowstone Fires of 1988," *BioScience* 39 (1989):678–85; Paul Schullery, "The Fires and Fire Policy," *BioScience* 39 (1989): 686–94; William Romme and Don Despain, "Historical Perspective on the Yellowstone Fires of 1988," *BioScience* 39 (1989):695–99.

26. Christensen et al., "Interpreting the Yellowstone Fires of 1988," 678.

27. Schullery, "The Fires and Fire Policy," 686.

28. Christensen et al., "Interpreting the Yellowstone Fires of 1988," 685.

29. Walters and Holling, "Large-scale Management Experiments," 2060.

30. Charles Birkeland, e-mail to the author, April 24, 1996.

31. International Coral Reef Initative, "Report to the United Nations Commission on Sustainable Development," 1996, 2.

INDEX